Lecture Notes in Mathematics 1521

Editors:
A. Dold, Heidelberg
B. Eckmann, Zürich
F. Takens, Groningen

Subseries:
Fondazione C. I. M. E.

Adviser:
Roberto Conti

S. Busenberg   B. Forte   H. K. Kuiken

Mathematical Modelling of Industrial Processes

Lectures given at the 3rd Session of the
Centro Internazionale Matematico Estivo (C.I.M.E.)
held in Bari, Italy, Sept. 24-29, 1990

Editors: V. Capasso, A. Fasano

Springer-Verlag
Berlin Heidelberg New York
London Paris Tokyo
Hong Kong Barcelona
Budapest

Authors

Stavros Busenberg
Department of Mathematics
Harvey Mudd College
Claremont, CA 91711, USA

Bruno Forte
Department of Applied Mathematics
University of Waterloo
Waterloo, Ontario, Canada N2L 3G1

Hendrik K. Kuiken
Philips Research Laboratories
P. O. Box 80000
5600 JA Eindhoven, The Netherlands

Editors

Vincenzo Capasso
Dipartimento di Matematica
Università di Milano
Via C. Saldini 50
I-20133 Milano, Italy

Antonio Fasano
Dipartimento di Matematica "U. DINI"
Università di Firenze
Viale Morgagni 67/A
I-50134 Firenze, Italy

Mathematics Subject Classification (1991): 00A71, 34A55, 35R30, 35R35, 35K55, 80A22, 80A23, 80A30, 82C32, 93A30

ISBN 3-540-55595-1 Springer-Verlag Berlin Heidelberg New York
ISBN 0-387-55595-1 Springer-Verlag New York Berlin Heidelberg

© Springer-Verlag Berlin Heidelberg 1992
Printed in Germany

Typesetting: Camera ready by author/editor
Printing and binding: Druckhaus Beltz, Hemsbach/Bergstr.
46/3140-543210 - Printed on acid-free paper

PREFACE

The role that university mathematicians can play in solving problems from real life has been the subject of discussions, round tables and symposia which are becoming more and more frequent. Certainly this matter does not belong specifically to our days: there are so many instances of illustrious mathematicians of the past who have used or even created mathematical tools for investigating problems of engineering, physics, chemistry, biology etc. The reason for such growing interest lies on the one hand in the increasing complexity of technological problems which require more mathematics, and on the other hand in the fact that university mathematicians are gradually rediscovering applied sciences as an inexhaustible source of appealing and challenging mathematical problems.

This set of CIME Courses demonstrated effectively such traits of so-called Industrial Mathematics. The lecturers are well-known mathematicians with a very large experience in the application of mathematics to problems submitted by industrial companies.

They described a variety of problems arising in different fields. All of them being brilliant teachers, their lectures were highly stimulating and very appropriate to illustrate one fundamental point: Industrial Mathematics IS FIRST OF ALL MATHEMATICS, not just modelling (although the role of mathematicians is often crucial in this stage), not just computing (usually the final stage of such kind of research), despite the fact that often a theorem has to be transformed into innovative software in order to become the terminal product "sold" to the company.

We are proud to have organised this CIME Session and we thank the Director and the staff of CIME for their support, and TECNOPOLIS for having offered the lecture room as well as financial and logistic help.

We are extremely grateful to the lecturers for their efforts in selecting the most appropriate material and for drawing such a clear picture of what Industrial Mathematics is today.

Vincenzo Capasso   Antonio Fasano

TABLE OF CONTENTS

Mathematical Modelling of Industrial Processes

H.K. Kuiken
Philips Research Laboratories
P.O. Box 80 000, 5600 JA Eindhoven
the Netherlands

Contents

Industrial Mathematics

What is industrial mathematics?

I am not sure who was the first to come up with the term industrial mathematics. The first time I heard it was during a visit to Bell Labs in Murray Hill in the summer of 1974. People like Henry Pollak of Bell Labs or Murray Klamkin of the Ford Motor Company used it freely in various publications. It is possible that the term was already used by Fry who headed the Math group of Bell Labs before WW II. Of course, the 'I' in the acronym SIAM stands for industry. Yet, most members of that society are applied mathematicians in the sense I shall define later on, rather than industrial mathematicians. When I subsequently mentioned the term in Europe, people looked at me in disbelief and dismissed it as an objectionable americanism. However, as happens over and over again with neologisms that seem to meet an existing need, the term was eventually widely accepted, also in Europe. Many universities have since created chairs of industrial mathematics. Societies such as ECMI (European Consortium for Mathematics in Industry) have been founded which profess their devotion to the application of mathematics in industry.

Hearing the expression 'industrial mathematics' for the first time, one can justifiably ask oneself whether this means anything different from the mathematics people use in non-industrial circles. In general, one should say, mathematics, whether this be topology or numerical analysis, does not depend upon the place where it is practised. The response of the industrial mathematician will be that the term does not specify a novel kind of mathematics, but rather emphasizes the fact that the application of mathematics in non-mathematical disciplines is the central theme. First, there is a problem arising outside mathematics and then whatever mathematics is needed is brought to bear upon it. Here mathematics is not a goal in itself, but rather a means to get answers in a non-mathematical world. Industrial labs such as Bell Labs in America or Philips Research Labs in Europe and, of course, there are many many more, are devoted to fundamental and basic research in many non-mathematical areas. Nevertheless, this makes them into veritable feeding grounds for the application of all kinds of mathematics.

Of course, the term 'applied mathematics' has been known to us for a much longer time and there seems to be good reason to question the necessity of introducing yet another expression for an activity which seems no different from ordinary applied mathematics. I feel that an explanation can be given when we consider the historical development of mathematics. Mathematics came into being because man felt the need to better govern the world around him and to improve his ability to predict phenomena that determined his life. Geometry was created in ancient Egypt for people to be able to cope with the problem of finding again each plot of land after the annual flooding of the Nile. The ancients of Mesopotamia were masters at devising tables with which they could predict conjunctions of the planets which, so they believed, determined the fate of nations. We shall not elaborate on this issue here. It should be clear, however, that these early, anonymous mathematicians were also physicists and engineers *avant la lettre*.

It would seem that this situation persisted until quite late in the nineteenth century. Great mathematicians such as Newton or Euler relied heavily on what they saw in the world around them for their mathematical inspirations. Indeed, it is very difficult to distinguish between Newton the Physicist and Newton the Mathematician. Mathematicians regard him as one of their leaders, because he is the (co-)inventor of differential calculus. Physicists call him their greatest on account of his three famous laws of mechanics. His *Principia* is good reading for physicist and mathematician alike. Only late in the nineteenth century do we see the emergence of a breed of investigators that we now call pure mathematicians. These people seem to be interested solely in the axiomatic worlds they themselves created. For their inspiration they do not rely on the world in which they live, but only on the world in which they think. This development ultimately leads to the likes of G.H. Hardy, who complimented himself on never having achieved anything of practical value, and Paul Halmos who tried to convince an audience that any mathematics that can be applied is dirty or ugly by definition. Quite amusingly, the fact of the matter is that most, and perhaps all, mathematics such people create is or will be used eventually by others to elucidate real-world problems. Sometimes this happens so fast that they live to see it.

This development of mathematics from applied to pure, which meant a turning away from the real world, an introspective move so to speak, did not mean that the application of mathematics got into low gear. On the contrary, many fields of science which had always had a verbal character to them, meaning that their truths were presented in plain language, were rapidly mathematized. However, people who called themselves mathematicians seemed to lose touch with this development. This game is now played by the physicists, the chemists, the engineers themselves. In the meantime, the mathematicians, at least the purer ones among them, are dreaming away in a world of their own making.

The central tenet of pure mathematics is formal proof. The sentiment about formal proof has become so strong among pure mathematicians that many of them will not accept as true any mathematics which has not stood this test. The difficulty is that most of the problems which arise outside mathematics are so complicated that it is impossible, at least for the time being, to give formal proof. As an example we could mention the mathematics that is used to describe flows around aeroplanes. Long before a real aeroplane is actually flown, its mathematical model has already been tested to the full. It is unthinkable that any modern aeroplane could be developed without the aid of extensive mathematical modelling. Yet, the correctness of these models has not been fully established. Even so, mathematicians, both pure and applied, use these machines for their long-distance travels. Only a handful of them will question the unproven safety these vehicles afford.

Who then are the inheritors in our day and age of the long tradition that began with our anonymous friends of the Nile and the Euphrates, all the way to Newton, the Bernoulli's, Euler, etc.? Indeed, there are such contemporaries, but if there had been a Nobel prize for Mathematics, which there is not, thanks to the strained relations between Nobel and Mittag-Leffler, then these people would not be among the laureates, for the simple reason that they would not be recognized as mathematicians. I mean people such as Theodore von Kármán and Geoffrey Taylor. The first of these applied his mathematical genious to almost every branch of engineering science. The same is true for Taylor who also had a knack for doing beautiful experiments and making simple but very elucidating mathematical models for them. We could also think of someone like John von Neumann. Pure mathematicians also regard him as one of their heros. Contrary to Hardy or Halmos, he also used his enormous talent to clarify problems emanating from other disciplines.

Again, I pose the question: is industrial mathematics any different from applied mathematics? No, not if we mean applied mathematics in the classical sense, linked with names such as Newton or Taylor. Of course, in the old days there were no industries to speak of, at least not ones that needed mathematics on a large scale. Modern industries are simply concentrated worlds within our world, where all kinds of problems arise that mathematics may be successfully applied to. Archimedes, Huygens, Newton would probably have taken a great interest in them. In principle, the manner in which mathematics is applied in these industries is not different from the way the classics operated. The difficulty with the term applied mathematics is that it is used nowadays to signify a different activity. Someone who studies properties of differential equations, although he/she may never solve a real-world problem, is called an applied mathematician, for the simple reason that differential equations are used to tackle problems arising outside mathematics. The same is true for someone interested in operator theory, Lie groups or numerical analysis. Clearly, in this sense applied mathematics is concerned with problems that arise within the mathematical world itself. It goes without saying that the fruits of this type of research are very useful for those mathematicians who tackle real-world problems.

Henry Pollak of Bell Labs distinguishes five stages in any separate activity of an industrial mathematical nature, which I shall repeat here in my own phraseology.

- Stage 1, *a problem arises outside mathematics.*
  This is the interface of mathematics and the real world. It often happens that the mathematician working in industry is approached by one of his/her non-mathematical colleagues who puts a technical problem before him/her. By problem I do not mean that the person in question wants to solve a differential equation or evaluate an integral. Of course, occasionally the industrial mathematician may help out with problems such as these, but this is not what he is in industry for. No, by technical problem I mean that the colleague describes a phenomenon or an experiment, and that he/she feels that the situation may be clarified by bringing mathematics to bear upon it. During this first encounter the discussion will be conducted mostly in ordinary language, although it may be interspersed with the jargon of the field the technical problem arose from. Depending upon the mathematical skills of the one who brought the problem, the discussion may take a mathematical twist. It should be clear that one of the prerequisites for this encounter to lead to any success or progress is that the industrial mathematician must be able to understand the language of the person who consults him/her. If his/her head is always high in the mathematical clouds, he/she will probably fail as an industrial mathematician. The successful industrial mathematician must therefore be interested in at least a few disciplines other than mathematics itself.

- Stage 2, *mathematical modelling.*
  This is where the original technical problem is cast into mathematical terms. I shall devote a separate section to describing what I think a mathematical model is all about. Let me state here that there is not one single mathematical model for each particular technical problem, but that we have a whole hierarchy of them. Aris describes this at length.

- Stage 3, *the analysis of the problem.*
  The purpose of the model is to obtain understanding. This understanding is obtained by analysis. These days, with all the computing power available, many industrial mathematicians are tempted to take the model to the computer as fast as they can, churning out numbers in endless series. Because these computers are so powerful, there is a tendency to devise models of ever-increasing complexity. The world models developed by economists are good examples of where this trend may lead us. Many hundreds or even thousands of parameters can be varied simultaneously, yielding a multitude of different answers. Although it is possible that this

is sometimes the only realistic way to proceed, I should like to state here that a professional industrial mathematician ought to postpone the use of the computer as long as he/she can. First, he/she should apply all the mathematical craft he/she can muster to work with the model, to modify it, to reduce it, to simplify it. If the primary model is dimensional, render it dimensionless. Then look at the dimensionless parameters. Are they small or large? Can the model be reduced by means of asymptotic methods? If the model is too complicated, can a simpler model give some clues? And so on and so forth.

- Stage 4, *numerical evaluation of results.*
  Except in the simplest of cases most models will eventually require the use of computers. This can be a simple matter of evaluating the value of an integral or an infinite series, or it can be a very complicated exercise in finite elements. In any case, the problem that is taken to the computer should be well balanced and computing time should be used economically. For instance, if someone has to calculate the value of a slowly-converging series which needs the evaluation of, say, one million terms to reach four-digit accuracy, then he/she has a very lazy mind if he/she writes a direct algorithm. First, one has to investigate whether convergence can be accelerated.

- Stage 5, *communication.*
  Since the problem came from outside, the results have to be returned to the outside world in a way that can be understood by outsiders. This aspect is neglected in many works which pretend to be applied or even industrial mathematics. Most mathematicians are happy and content when they have solved a problem their way, meaning that they have finally produced an intricate formula, a nice algorithm or splendid proof. However, this language is rarely understood by non-mathematicians. The successful industrial mathematician, on the contrary, will spend a great deal of his/her energy on making clear graphs, fine tables and a lucid verbal description of the results obtained. An accomplished industrial-mathematics report will contain a clear description put in ordinary language of the problem to be tackled. Then follows a description of the model and a sketch of the methods that were used to obtain the solution. Extensive calculations which would not interest anyone but the real experts are best reported in appendices. The report ends with one or two sections on results and a discussion of the results. Again the emphasis should be on the use of ordinary plain language.

What is a mathematical model?

A mathematical model is an imperfect image of a part of the world around us, in which use is made of mathematical symbolism. The model employs mathematical representations of the basic laws of nature, for instance the conservation laws. In this respect it may be useful to remark that the word 'nature' is to be interpreted in the widest sense possible. It is not restricted to what physicists understand by nature, but it includes also things abstract; to put it shortly, everything our brain can grasp. As a special feature, a mathematical model, despite its imperfections, provides us with an insight into those parts of the 'world' which are inaccessible to us, either for the time being or permanently. Mathematical models have been made to describe conditions in the core of the sun or the earth. Clearly, we shall probably never have bodily access to those parts of the world, and neither will it be possible, at least for the time being, to carry out experiments there. Nevertheless, we feel that these models enable us to predict what happens in those remote places. Mathematical models are also helpful in predicting things which are still to happen, to predict future occurrences, so to speak. According to a set of assumptions, different scenarios are presented to politicians, managers, the military, whose decision on what course of action is to be taken depends upon what the models foretell.

We have already pointed out the imperfection of mathematical models. This has to do with the fact that in building such a model we can represent only a very limited number of aspects of the part of the world we wish to know more about. Depending upon whether our approach is cruder or less crude, we may refer to our models as very imperfect or less imperfect. The euphemism 'refined' is sometimes used for models which, in reality, are only less imperfect. However this may be, it should be clear that each part of our world can be modelled in a great many different ways. In each particular instance we can distinguish a complete hierarchy of mathematical models from crude to refined. The level of refinement we can reach may depend upon the time we can spare, upon our financial means and, last but not least, upon the limitations of our brain power.

Apart from mathematical models we also know physical models. Whereas with a mathematical model we are concerned with an abstract representation of some part of our world, be it abstract or tangible, a physical model is a tangible representation of a tangible part of the world. That particular tangible part of the world is copied on a reduced and simplified scale. Architects use such models. Civil engineers have a great tradition in this particular area. They have earned themselves fame with their physical models of tidal systems for low-lying coastal areas. Sometimes the sizes of these models are gigantic, occupying many thousands of square metres. Nevertheless, despite the great sums that are spent on these physical models, they do have very definite drawbacks. The most important of these is that it is never possible to scale down all variables of the modelled system in the same fashion. For instance, for practical reasons the rivers in the aforesaid tidal models are always much deeper than they ought to be. If these depths were modelled to scale, then the surface tension of the water would affect the working of the model in an intolerable fashion. Scaling up and scaling down are very tricky operations. It is precisely through the study of mathematical modelling that we know that a scaled-down version of a physical reality can be in a theoretical regime which is completely different from that of the physical reality itself. Of course, nature itself has known this all along: There is no such thing as the one-millimetre elephant, nor is there, contrary to what some science-fiction movies would have us believe, room in our world for the one-metre ant. Be this as it may, it would seem that physical models are things of the past. Yet, with man's disposition being such that, on the whole, he is inclined to favour things tangible, these physical models are likely to remain with us for a long time.

We remarked earlier that in making a mathematical model we adopt a selection process in which only certain attributes of the world around us are seen fit to be represented in the model, whereas others are left out of consideration. For the latter we use the word 'neglected'. It often happens that after the model has been set up, the process of neglecting certain effects is carried still further on the basis of a process of mathematical reasoning conducted within the model. Although after the making of the model, physical reality is represented by a purely mathematical object which, strictly speaking, we can talk about in mathematical terms only, it can sometimes be helpful to refer back to the physical reality the model is thought to be an image of. We have to be careful, however, not to refer to physical entities that are not represented in the model. This would seem to be a matter of course. However, the following anecdote serves to show that confusion can arise if one is not careful in one's reasoning:

A mathematician working in industry had made a model for the diffusion of electrons in a layer of a few thousand ångströms. In his model he made use of the well-known diffusion equation for continuous media. This would seem a reasonable thing to do for a layer that was many thousands of atoms thick. He presented the fruits of his research to an audience that was composed mainly of physicists. He told them that he had used finite differences to solve the problem and that he had found it to be necessary to divide the layer into many thousands of subintervals in order to attain the required accuracy. Someone in the audience questioned the validity of the model, since each subinterval was of subatomic length. Continuous models are not valid then. This remark baffled the speaker, who could not come up with a convincing answer. The audience left the lecture room, being more convinced than ever that these mathematicians had nothing useful to offer. Of course, after a moment's thought, the correct answer is easily found. The subdivision of the interval occurs *within* the mathematical model. Although the subintervals are deceptively like atoms, they have nothing to do with them. If the physicist agrees that a continuous model is valid when the modelled physical entity contains many thousands of atoms, then a mathematician can prove to him that the solution to that valid model will be approximated more and more accurately when he applies finer and finer meshes.

Finally we should be careful not to overestimate the power of a model and not to be too absolute about the results derived from it. The model is as good or as bad as the assumptions on which it is founded. If it is at all possible, one should test its validity by means of experiments. In the event that the quality of the results is inadequate, one might consider refining the model. Of course, all depends upon one's goals. If one's goal was the procurement of qualitative insight, for instance about the direction in which a process will run or about the order of magnitude of certain effects, then a crude model may do. If one needs numerical answers that have a certain measure of accuracy, then one may opt for mathematical models that are of the less imperfect kind, refined if you like. A report which appeared in the Dutch newspaper NRC Handelsblad on October 29, 1986, is quite illustrative in this respect. The newspaper article dealt with some unexpected flow levels in a newly dug canal in the south-west of the Netherlands: I shall present an English translation of the Dutch original:

Shipping in the Rhine-Scheldt canal is severely hampered by unexpectedly high flow velocities in this canal. These result from the closing last week of the Oesterdam (Oyster dam) between Zuid-Beveland and Tholen ...

... Mr. Hamer (M. Eng.) of the Ministry of Public Works said this morning that computer models have led him to expect that the flow velocities would have increased from 0.5

metres per second to 1.5 metres per second. Now this has turned out to be 2 metres per second. This can cause problems, since the transport of soil from the canal bed increases disproportionately. In the calculations use was made of both a physical model and a one-dimensional mathematical model. Hamer felt that it would be unjustified to state that the models were flawed or that nature had played yet another trick on technology. "We shall proceed by organizing all our data and making the right comparisons. Only then shall we be able to draw the right conclusions", he said.

His remarks about the surprises nature has in store for us show that Mr. Hamer seriously overestimated the power of his mathematical models. Of course, it is quite inconceivable that one should be able to simulate accurately a flow system as complicated as the Rhine-Scheldt delta by means of a one-dimensional model. Qualitative answers are the best one can expect from such a simple model. On the contrary, it is surprising that the Public-Works people, using their simple model, should have been able not only to predict the direction in which the flow field would change, but also two-thirds of its magnitude. This is ample evidence of the expertise these people have acquired over the years.

What about a lecture series on industrial mathematics?

I have been asked to say a few things here in Bari about the application of mathematics in industry, so as to give the audience an idea of what industrial mathematics is all about. From what I have said up to now it should have become clear that this is an impossible task. In principle, the subject encompasses almost everything the human mind can think of. It is not for one person to present an overview of such a wide field, and certainly not if he is allowed only one week to do it in. All I can do is talk about a few things that have come my way. Next year when you invite someone else, you will hear a completely different story. But then, looking at one painting will not make one an expert on painting, nor will reading one poem transform the reader into a master of poetry. Of course, if you never look at a painting or read a poem, ...

Suggestions for further reading

- R. Aris, *Mathematical modelling techniques*, Pitman, 1978.

- W.E. Boyce (Editor), *Case studies in mathematical modeling*, Pitman, 1981.

- M.S. Klamkin, On the ideal role of an industrial mathematician and its educational implications. Educ. Stud. in Math. **3** 244-269, 1971.

- C.C. Lin and L.A. Segel, *Mathematics applied to deterministic problems in the natural sciences*, Macmillan, 1974.

- H.O. Pollak, How can we teach applications of mathematics? Educ. Stud. in Math. **2** 393-404, 1969.

- A.B. Tayler, *Mathematical models in applied mechanics*, Clarendon, 1986.

Some relevant journals

- Journal of Engineering Mathematics (quarterly)

- Mathematical and Computer Modelling (monthly)

- Mathematical Engineering in Industry (quarterly)

- SIAM Journal of Applied Mathematics (bi-monthly)

Temperature distribution within
a crystal-growing furnace

Motivation

The purpose of this chapter is to illustrate the process of reduction, which is an important element in the development of a mathematical model. The mathematical model is an image of a part of the world around us. If this part is in itself rather complicated, then a truthful representation of it in mathematical terms means that we must consider an equally complicated mathematical model. This is not always what we want. Therefore, as occurs frequently when making a mathematical model, we follow a process of reduction. The final model is always a compromise between what we would like to achieve and what we can actually realize.

Discussion of the technical background

Crystals are very important semi-manufactured products in the electronics industry. For instance, silicon crystals are used as a basic material for the manufacture of chips. Materials such as gallium arsenide (GaAs) and cadmium telluride (CdTe) are applied in lasers. The basic property of crystals which makes them so desirable is that their atoms are ordered. The way the atoms are ordered depends upon the material in question. We distinguish cubic, hexagonal, rhombic, etc., crystallographic structures. Quite frequently, industry produces large so-called single crystals, the sizes of which can range from a few to many tens of centimetres. In such a single crystal the basic structure and orientation are the same throughout the crystal. The aforementioned products are made by cutting a large single crystal into smaller parts (chips). A basic problem is that these large single crystals are hardly ever perfect. The ideal crystallographic structure is interrupted by faults which are called defects. These defects are created during the crystal-growing process, owing to induced stress levels. Mostly, thermal stresses are to be blamed. These defects often show a tendency to propagate themselves by diffusion throughout the crystal, thereby completely spoiling it. It would seem, therefore, that temperature control is of the utmost importance in the aforesaid production processes.

A few methods are available for the production of these large, mostly cylindrical, single crystals. One of these is the so-called Bridgman-Stockbarger technique. In this process a powdery mixture of the materials the crystal is to be composed of is put into a cylindrical shell or container, which is then closed. This container or crucible is then slowly moved downwards through a vertically positioned furnace. Within this furnace there is a special hot zone in which the temperature is high enough to melt the powder. On leaving the hot zone, the melt recrystallizes, hopefully as the desired perfect single crystal. Some crystal growers favour a narrow but intense hot zone, others an elongated mild one. The present study arose from precisely this conflict.

One of the problems facing the crystal grower is that it is very difficult or even impossible, because of the extremely high temperatures, to obtain direct experimental insight

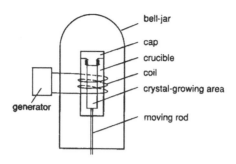

Figure 1: Sketch of crystal-growth system.

into what happens in the hot zone. Moreover, the processes of melting and recrystallization occur within a closed container. Nevertheless, it is still very important for the crystal
grower to know the isotherm structure around the solid-liquid interface. A highly curved
s-l interface may give rise to induced thermal stresses and hence to the creation of defects.
The ideal situation is one with a flat interface. The question is how to bring this about.
Often different ways of heating will be necessary to achieve the ideal s-l interface shape
during the various stages of the crystal-growing process.

It should be clear that a mathematical model may offer a way out where experimental
methods fail. Our purpose will be to write down a set of equations which govern heat
transfer around and inside the crucible. Moreover, we shall have to describe boundary
conditions which determine the exchange of heat with the surrounding world. It will not
come as a surprise that, because of both the complicated geometry and a multitude of
phenomena that are physically relevant, this may give rise to models that are extremely
complicated. We must therefore ask ourselves right at the beginning of our enterprise
whether a broad approach is really what we want, assuming of course that we shall be
able to formulate and then solve an all-encompassing model. The average crystal grower
is often simply groping about in the dark concerning even the most basic aspects of his
process. Therefore, models that provide insight will be most welcome. The maker of a
suitable mathematical model therefore faces the task of reducing the problem definition
to such an extent that the ensuing model will be manageable, without trivializing the
subject matter in an intolerable fashion.

Let us now put these principles to the test, using a Bridgman-Stockbarger configuration with a narrow heating zone. Fig. 1 gives a rough sketch of the furnace system. The
crucible is in the centre of the furnace. During the crystal-growing process it moves in
a vertical direction, mostly downwards. Sizes of crucibles are from ten to a few tens of
centimetres. Widths may range from two to five centimetres. Ideally a crucible is quite a
bit longer than it is wide. Surrounding the crucible is a co-called r-f coil. This is a hollow
curled-up tube, mostly made of copper, with a diameter of about half a centimetre. Water
flows through the tube for cooling purposes. The coil is connected to an electric element
which produces a high-frequency current within it. This alternating current in turn pro-

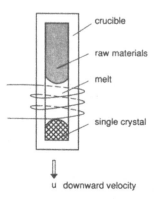

Figure 2: Simplified model of crucible.

duces a strong electromagnetic field in the immediate vicinity of the coil. If the crucible is composed of a suitable material such as graphite, the field will produce intense eddy currents within it. In the majority of cases the penetration depth of the field is limited, so that the currents are strong only in the part of the crucible that is close to the coil. The eddy currents cause dissipation (Joule heating), thereby heating up the crucible and melting the powder inside. When the crucible is moved downwards through the coil, we first see a melting of the powder just above or in the upper reaches of the coil, followed by a recrystallization, hopefully as a perfect single crystal, below the coil (Fig. 2).

The crucible and the coil are completely surrounded by a perfectly sealed bell-shaped structure. It is often necessary to prevent certain gases, e.g. oxygen, from coming into contact with the crucible. Not infrequently the pressure within the bell-jar may exceed the atmospheric pressure many times. The reason is that the vapour pressure of the molten crystal material increases rapidly with temperature. When it is perfectly sealed, this may give rise to high pressures within the crucible. In the case of an imperfect seal, volatilization may occur. A high pressure within the bell-jar may compensate for these detrimental effects.

It would seem that in making a mathematical model we should at least consider the following effects and phenomena:

1. Generation of the electromagnetic field and the manner in which this field gives rise to eddy currents within the graphite the crucible is made of;

2. The production of Joule heat by these eddy currents;

3. The conduction of heat through the crucible to the charge (powder + melt + crystal), then through the charge, and finally towards the colder parts of the crucible away from the coil.

4. Heat transfer from the outer surface of the crucible to the surroundings. This occurs through radiation and convection.

Apart from these principal items, other phenomena, such as latent heat and heat transfer away from the outer surface of the bell-jar, may have to be considered.

Clearly the list of problems quoted will probably baffle even the most optimistic of mathematicians. Yet, it is true that all items in the list will have to be represented accurately, if the model is to be predictive of the actual experimental situation. As we have already emphasized on several occasions, the experimenter may be quite happy to settle for a far less ambitious goal. Of course, he would like to have a fairly good idea of what happens within the crucible. Therefore, our willingness to compromise can only be limited in the case of item three on the list. The remaining three points cause severe difficulties. A complicated geometry (coil, bell-jar wall) seriously hampers the modelling of radiation transfer. To describe convective heat transfer in the gas, complicated nonlinear equations are needed. The complete set of Maxwell's equations will be necessary for the description of the transfer of energy from the coil to the crucible. These problems would all disappear if we knew the temperature at the outer surface of the crucible.

In view of what was said above, when developing the first in a series of models for this system, it would seem to be a good idea for us to assume the temperature known on the outside of the crucible. Such an assumption is not at all unrealistic. Using a set of thermo-couples or a pyrometer, the experimenter will be able to record the required temperature profile at any point in time. The temperature being known at the outer surface of a body, a field equation will determine the temperature elsewhere within that same body. We are no longer required to tell how the heat is transferred to and away from the body. This is an enormous simplification which renders the problem at once tractable.

The problem definition is now reduced to one concerning heat transfer within a finite, elongated, slowly moving cylindrical body that is composed of a number of different materials. Such a problem is solvable in a practical sense using numerical methods. We shall simplify the problem even further. As we said before, Bridgman-Stockbarger crucibles are usually fairly slender, the reason being that a large part of the crystal is then grown under the same conditions. End regions cause non-uniformities. In short crucibles these would be too prominent. Therefore, in our initial model the crucible will be infinitely long. We shall also treat the powder and the crystal as thermally similar materials. This is incorrect, of course. The next model in the hierarchy of models could be more precise on this score.

Fig. 3 shows the geometry of our reduced initial model. The reasoning used in the reduction process was of a purely verbal character. No mathematics has played a role yet. We shall now proceed by mathematizing our model and reduce it even further. However, those reductions will now be based on mathematical reasoning. In Fig. 3 we give a cylindrical coordinate system fixed in space. The origin of the axial coordinate z has been taken in the plane where the wall temperature $T_w(z)$ reaches its maximum value T_M. Assuming that T_M is high enough, we can expect solid-liquid interfaces on both sides of the plane $z = 0$. The cylinder moves at a velocity u in the negative z-direction.

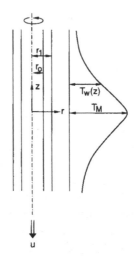

Figure 3: Given temperature profile.

Setting up a mathematical model

Steady heat transfer in a moving medium is governed by a convection-diffusion equation

$$\rho c_p \mathbf{v}.\nabla T = \nabla(\lambda \nabla T) \ , \tag{1}$$

where T denotes the temperature. Further ∇ is the gradient operator. The density is denoted by ρ, the specific heat by c_p and the thermal conductivity by λ. Assuming that there is no convection in the melt, we conclude that the cylinder as a whole moves as a rigid body, so that the velocity vector can be written as follows

$$\mathbf{v} = (0, 0, -u) \tag{2}$$

in a (r, φ, z) coordinate system. We can therefore write (1) as follows

$$\frac{1}{r}\frac{\partial}{\partial r}\kappa r\frac{\partial T}{\partial r} + \frac{\partial}{\partial z}\kappa\frac{\partial T}{\partial z} + u\frac{\partial T}{\partial z} = 0 \ , \tag{3}$$

where κ is the thermal diffusivity

$$\kappa = \frac{\lambda}{\rho c_p} \ , \tag{4}$$

which is a function of temperature, since λ is a function of temperature. Both ρ and c_p are assumed to be independent of the temperature.

Before going on, it will be useful for us to keep track of the dimensions of the various parameters and variables we defined up to now. We shall always express these dimensions in terms of the MKSA system. Using [ ] to denote the dimension, we have

$$[T] = \ K; \ \ [u] = m.s^{-1}; \ \ [\rho] = kg.m^{-3}; \ \ [c_p] = J.kg^{-1}.K^{-1};$$

$$[\nabla T] = K.m^{-1}; \quad [\lambda] = J.m^{-1}.s^{-1}.K^{-1}; \quad [\kappa] = m^2.s^{-1} \text{ etc.} \tag{5}$$

It is also useful to define the heat-flux density

$$\mathbf{q}'' = -\lambda \nabla T \ , \tag{6}$$

which has the dimension

$$[\mathbf{q}''] = J.m^{-2}.s^{-1} \ . \tag{7}$$

Eq. (7) tells us that the inner product of the vector $\mathbf{q}''$ with a unit normal $\mathbf{n}$ on an arbitrary plane is equal to the amount of heat flowing through a unit area on that plane per unit of time.

Apart from the quantities defined above we have to consider the inner and outer radii of the crucible, which are denoted by r_0 and r_1, respectively. Furthermore, there is the maximum temperature T_M on the outer wall of the crucible and a length l which characterizes the longitudinal variation of the wall temperature $T_w(z)$:

$$T_w(z) = T_M f(z/l) \ , \tag{8}$$

where f is a dimensionless function which has a maximum value $f(0) = 1$. We must also consider the temperature T_t at which the crystal melts. It should be clear that this fairly simple problem contains already a large number of independent parameters, viz. $r_0, r_1, u, T_M, T_t, \lambda, \rho, c_p, l$. This observation applies even more forcibly, since some of these parameters assume different values in each of the three materials occurring in the system: crucible material, melt and crystal.

Although the temperature has been described at the outer boundary of the system, namely by Eq. (8) to which we add

$$T \rightarrow T_M f(\pm\infty) \quad \text{if} \quad z \rightarrow \pm\infty \ , \tag{9}$$

we need to consider conditions at the internal interfaces. At each of these interfaces we require continuity of the temperature. There are interfaces in nature across which the temperature jumps, but we shall not consider these here. If no heat is lost or produced at an interface, the heat flux $\mathbf{n}.\mathbf{q}''$ must be continuous also. We see from (6) that this means the continuity of $\lambda \partial T / \partial n$ across the interface. Since λ assumes different values on both sides of the interface, the normal derivative of T is discontinuous.

There is still a matter of notation to be discussed here, which usually does not arise in a mathematical discussion. In an expression such as $\mathbf{n}.\mathbf{q}''$ the symbol $\mathbf{n}$ denotes a dimensionless quantity, namely a unit vector. In the term $\partial T / \partial n$, on the other hand, the symbol n has the dimension of a length. One should keep this in mind when one decides to render a problem dimensionless.

At the interface between the crucible material and the charge (crystal + melt), which is defined by $r = r_0$, we have

$$T(r_0 - 0, z) = T(r_0 + 0, z) \ , \tag{10}$$

$$\lambda_b \frac{\partial T}{\partial r}(r_0 - 0, z) = \lambda_a \frac{\partial T}{\partial r}(r_0 + 0, z) \ , \tag{11}$$

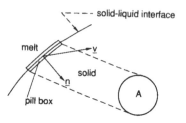

Figure 4: Pill Box.

where λ_a is the (constant) value of λ of the crucible material. By λ_b we denote the thermal conductivity of the charge. This quantity assumes different values in the melt and the crystal:

$$\lambda_b = \begin{array}{ll} \lambda_m & (T > T_t) \\ \lambda_c & (T < T_t) \end{array} . \tag{12}$$

Again both λ_m and λ_c are assumed constant.

The interfaces between the melt and the crystal are slightly more difficult to deal with. First of all, it is not known *a priori* where they are located. Indeed the main object of our model is to tell what the solid-liquid interfaces look like, i.e. where they are located. Let us denote the s-l interface by Γ. Then on Γ we have

$$T \mid_c = T \mid_m = T_t \ , \tag{13}$$

which means that the temperature on both sides of the s-l interface is equal to the transition or melting temperature T_t. Since we consider a steady problem here, Γ will be fixed in space. Since the cylinder moves, the material crossing Γ will change phase. It is well known from thermodynamics that this phase change is accompanied by the liberation or consumption of latent heat, depending upon whether the material solidifies or melts. This latent heat is denoted by L with

$$[L] = J.kg^{-1} \ . \tag{14}$$

This physical parameter tells us how much heat is released or taken up when a unit mass of the material in question changes phase. Fig. 4 shows a control volume, a so-called pill-box, which encloses a small part of Γ, with an area equal to A. The pill-box is assumed to be infinitely thin so that its width can be neglected in comparison with A. If $\mathbf{n}$ is the unit normal to Γ pointing into the solid, then the amount of heat liberated within the pill-box is equal to

$$A\mathbf{n}.\mathbf{v}\rho L \ , \tag{15}$$

where $\mathbf{v}$ is the vector velocity of the material. Strictly speaking (15) is valid only if A is infinitesimal. It is always useful in exercises of this kind to check if the dimensions come out right. From (5) and (14) we find J.s^{-1} as the dimension of (15), which is correct. We can now write down a thermal balance for the pill-box, namely that if we add the heat defined by (15) to the heat conducted into the pill-box from the melt, we must end up with the heat conducted into the solid, or in mathematical terms

$$A\mathbf{n}.\mathbf{v}\rho L + A\mathbf{n}.\mathbf{q}''\,|_m = A\mathbf{n}.\mathbf{q}''\,|_c \quad . \tag{16}$$

From (6) and (16) we find

$$\lambda_m \frac{\partial T}{\partial n}\,|_m - \lambda_c \frac{\partial T}{\partial n}\,|_c = \rho L \mathbf{n}.\mathbf{v} \quad \text{on} \quad \Gamma \quad . \tag{17}$$

As before, the meaning of the vertical bars is 'at the side of' m in the case of the melt and c when we mean the crystal.

In deriving (17), using Fig. 4, we adopted a 'physicist's approach'. Using macroscopic control volumes, mathematicians have devised methods that are far more elegant. However, the end result is the same, and that is what we are concerned with here.

A dimensionless formulation

We shall now make the model dimensionless. This is a necessary, even a central element within any activity where physical phenomena are modelled mathematically. There are two reasons why this process of rendering a model dimensionless is important. The first is that the number of independent parameters is reduced. The effect of this simplification is strongest when the number of parameters is already small to begin with, let us say five or six at the most. The so-called pi-theorem tells us that for each characteristic dimension a reduction of one parameter is possible. In our model we have three characteristic dimensions, viz. kg, K and m.

A second, and possibly far more important reason for embarking upon a course of non-dimensionalization is that we now get dimensionless parameters which fully characterize the problem by the values they have. Each parameter represents a particular aspect of the problem. If its value is around unity, then this particular aspect is important for the problem. If the parameter value is extreme, i.e. close to zero, let us say smaller than 0.1, or nearer infinity, let us say larger than 10, then the particular aspect of the problem it represents can be treated in an asymptotic sense. Its effect can either be disregarded to first order, or it is felt only in restricted areas, such as boundary layers. Through asymptotics the model can be reduced even further. This powerful tool is not only effective when the number of parameters is small, but also when they are many.

The chance that a particular dimensionless parameter is extreme is fairly large. We could even make a case for the observation that for a dimensionless parameter to be of order unity is sooner an exception than a rule. Indeed, each particular dimensionless parameter is obtained by a process of multiplying and dividing separate dimensional parameters. Given a particular dimensional system, these dimensional parameters can assume values which can vary through many orders of magnitude, depending upon the particular material or geometry which applies. As an example consider the thermal conductivity λ. For copper we have $\lambda = 400 J.m^{-1}.s^{-1}.K^{-1}$ and for air $\lambda = 0.025 J.m^{-1}.s^{-1}.K^{-1}$, revealing a difference of more than four orders of magnitude. Obviously, sizes of systems can also differ by many orders of magnitude. Now accepting this wide spectrum of values expressed in powers of ten, we are led to the conclusion that it is fairly improbable for a quantity which is composed of different elements, each of which can assume widely differing numerical values, to have a value around 10^0. This is why the one-millimeter elephant and the one-metre ant can only be created in the minds of film-makers who lack the necessary technical insight. Nature does not accept them, because they have got their dimensionless numbers wrong. Common fish are large-Reynolds-number creatures. They have got tails with which they can propel themselves. Microscopic aquatic creatures need a completely different way to get along. They are small- Reynolds-number animals that displace themselves by means of a motion which is more like that of a corkscrew. There is a beautiful science movie in which G.I. Taylor - yes, the one who would not have been eligible for the imaginary Nobel prize for mathematics[1] - explains these subtle differences.

After this digression let us now return to our main theme. We have to select two characteristic parameters with which we can render the spatial variables and the temperature dimensionless. Reasonable choices would seem to be r_1 and T_t, whence

$$r = r_1 R , \quad z = r_1 Z , \quad T(r,z) = T_t \theta(R,Z) , \tag{18}$$

where R, Z and θ are the dimensionless variables. Substituting this in the equations (3), (8), (9), (10), (11), (13) and (17), we obtain the folowing set of equations and boundary conditions:

$$\frac{1}{R}\frac{\partial}{\partial R}\Lambda R\frac{\partial \theta}{\partial R} + \frac{\partial}{\partial Z}\Lambda\frac{\partial \theta}{\partial Z} + \varepsilon\Lambda\frac{\partial \theta}{\partial Z} = 0 , \quad \begin{pmatrix} 0 < R < 1 \\ -\infty < Z < \infty \end{pmatrix} \tag{19}$$

$$\theta = \theta_M f(R_l Z) \quad \text{for } R = 1 , \tag{20}$$

$$\frac{\partial \theta}{\partial R}(R_0 + 0, Z) = \Lambda_b \frac{\partial \theta}{\partial R}(R_0 - 0, Z) , \tag{21}$$

$$\theta(R_0 + 0, Z) = \theta(R_0 - 0, Z) , \tag{22}$$

$$\frac{\partial \theta}{\partial n}\Big|_m - \Lambda_c \frac{\partial \theta}{\partial n}\Big|_c = G\cos\phi \quad \text{on } \Gamma , \tag{23}$$

$$\theta\big|_m = \theta\big|_c = 1 \quad \text{on } \Gamma , \tag{24}$$

$$\theta \to \theta_M f(\pm\infty) \quad \text{when } Z \to \pm\infty \quad (0 \le R \le 1) . \tag{25}$$

The parameter Λ is the dimensionless thermal conductivity. It is obtained by dividing the dimensional thermal conductivity by a reference thermal conductivity. For the latter we choose two values, one for the crucible, namely λ_a, and one for the charge, namely λ_m. Therefore

$$\Lambda = 1 \quad \text{for} \quad R_0 < R \le 1 , \tag{26}$$

$$\Lambda = \begin{cases} 1 & \text{in the melt } (\theta > 1) \\ \Lambda_c = \lambda_c/\lambda_m & \text{in the crystal } (\theta < 1) \end{cases} \quad (0 \le R < R_0) . \tag{27}$$

The parameter ε is given by

$$\varepsilon = \frac{\rho c u r_1}{\lambda_a} \quad \text{for } R_0 < R \le 1 , \quad \varepsilon = \frac{\rho c_p u r_1}{\lambda_m} \quad \text{for } 0 \le R < R_0 , \tag{28}$$

---

[1]see section entitled "Industrial Mathematics"

because of the different choices of the reference thermal conductivity. The parameter ε is truly representative of the relative magnitudes of the convective term and the conduction term, both in the crucible and the charge. Other parameters appearing in (19)-(25) are defined as follows

$$\theta_M = T_M/T_t \ , \quad R_l = r_l/r_1 \ , \quad R_0 = r_0/r_1 \ , \quad \Lambda_b = \lambda_b/\lambda_a \ . \tag{29}$$

Finally, we have the important parameter

$$G = \frac{u\rho L r_1}{\lambda_m T_t} \ . \tag{30}$$

Counting correctly, we find that the number of parameters which determine this dimensionless problem is equal to seven, viz. Λ_b, Λ_c, ε, θ_M, R_l, R_0 and G, although strictly speaking there are eight, since ε has two representations (see Eq. (28)).

Let us now make an estimate of the value of ε. The growth velocities of crystals being of the order of 1 cm per hour at the most, we can write $u \sim 10^{-6} m.s^{-1}$. For r_1 we have $r_1 \sim 10^{-2} m$. We still have to find a realistic value of $\kappa = \lambda/\rho c_p$ which is the thermal diffusivity. For a material such as copper we have a value $\kappa \sim 10^{-4} m^2.s^{-1}$. For water, which has a relatively large heat capacity, a value of $\kappa \sim 10^{-7} m^2.s^{-1}$ is given. Most other materials have values that are in between these two extremes. It follows that ε is a small dimensionless parameter in problems of the kind considered here. As a consequence we are allowed to neglect the term in which it appears as a multiplication factor. Physically speaking this means that the motion of the crucible has a negligible effect upon the temperature distribution. Of course, one must always be careful when one disregards terms which are multiplied by small parameters. One can only do this if the solution is an analytic function of the small parameter when its value approaches zero. In any case, choosing the correct scalings is an art that comes with experience.

Another parameter which deserves special attention is G (Eq. (30)). Let us see what value this parameter has when the charge consists of water and ice. In that case we have $u \sim 10^{-6} m.s^{-1}$, $\rho \sim 10^{-3} kg.m^3$, $L \sim 3.10^5 J.kg^{-1}$, $r_1 \sim 10^{-2} m$, $\lambda_m \sim 1 J.m^{-1}.s^{-1}.K^{-1}$, $T_t \sim 3.10^2 \ K$ so that $G \sim 10^{-2}$. For other materials similarly small values of G are found. Therefore, we are led to assume that the right-hand side of Eq. (23) can be put equal to zero. This simply means that the effect of latent heat can be neglected to first order.

The foregoing discussion has clearly shown what a process of rendering a problem dimensionless can do for us. It shows us the way to simplify the problem further and it reveals unambiguously the reasons why we can do it. The largest simplification was carried out when the original dimensional problem was formulated, because out of a multitude of effects and phenomena that could possibly play a role, we selected only a few for further consideration. A process of non-dimensionalization is then invoked to tell us which among these selected effects and phenomena are more prominent than others. It is an indispensable tool for any multi-phenomena study.

In our simplified model we now have to deal with the field equation

$$\frac{1}{R}\frac{\partial}{\partial R}\Lambda R\frac{\partial \theta}{\partial R} + \frac{\partial}{\partial Z}\Lambda\frac{\partial \theta}{\partial Z} = 0 \ , \tag{31}$$

where Λ is given by (26) and (27). The boundary and interface conditions to be satisfied are (20), (21), (22), (24) and (25) together with

$$\frac{\partial \theta}{\partial n} \mid_m = \Lambda_c \frac{\partial \theta}{\partial n} \mid_c \quad \text{on} \quad \Gamma \ . \tag{32}$$

This problem involves five dimensionless parameters, viz. Λ_c, Λ_b, R_0, R_l and θ_M. It is a non-linear, and therefore non-trivial, problem. The nonlinearity is caused by the fact that Γ is unknown and Λ changes abruptly across Γ.

Discussion about possible methods of solution

Now that we have finally formulated our model in its simplest form, we can start thinking about how to solve it. First of all, we must have a clear idea about what we want from the model. Although the temperature field is interesting, we are interested first and foremost in the shape of the liquid $\rightarrow$ solid interface, because this is the place where the crystal is actually grown.

A number of techniques are available to solve problems of the kind considered here. A fully analytical solution is probably unattainable, since the problem is nonlinear. Therefore, numerical techniques seem to be required. These days most people would opt for finite elements. However, we shall not do this here, the reason being that the shape of the unknown interface may have subtleties that a crude numerical approach may fail to bring to light.

Because of the large differences between the values of the thermal conductivities of the crucible material (graphite is a good heat conductor) and the charge (crystal and melt), a semi-analytical approach may still be possible. Most of the heat conduction will occur within the graphite. The heat enters the crucible in the high-temperature region and most of it is conducted through the crucible proper, i.e. the crucible system without the charge, to the low-temperature regions. Only a small amount of the heat will take the longer route through the charge. This offers a way of solving the problem iteratively using partial analytical solutions.

Because of what we just said, we may conclude that Λ_b is small. Therefore, Eq. (21) may be approximated to first order by

$$\frac{\partial \theta}{\partial R}(R_0 + 0, Z) = 0 \ . \tag{33}$$

Put in physical terms this means that no heat is flowing into the charge. Since $\Lambda = 1$ everywhere within the region $R_0 < R \leq 1$, we have formulated an easy-to-solve problem for the temperature distribution in the crucible proper. It consists of Eqs. (20), (25), (31) and (33) and can be solved analytically. Invoking condition (22) we find that this solution defines a temperature on the cylinder $R = R_0$, defining a problem in the region $0 \leq R < R_0$ consisting of Eqs. (24), (25), (31) and (32). Although, according to Eq.(27), the dimensionless thermal conductivity Λ changes abruptly across the unknown interface Γ, this partial problem can be solved analytically as well. This analytical solution defines a heat-flux distribution on the cylinder $R = R_0 - 0$ which, according to Eq. (21), can be transformed into a heat-flux distribution on the cylinder $R = R_0 + 0$ by multiplying it by the small parameter Λ_b. We can now solve a slightly modified heat-conduction problem in the annular region $R_0 < R \leq 1$, repeating the previous iteration steps until convergence to a stable solution is achieved.

Although the iteration process sketched above is based upon analytical solutions, we must still use numerical discretisation techniques. The temperature distributions on $R = 1$ and $R = R_0$, and also the heat-flux function on $R = R_0$, will have to be represented in a discretised manner. We shall not enter into the matter of proving the validity of the iteration process. In the Appendix we show how the method works when applied to a one-dimensional problem. That simplified case also suggests the usefulness of underrelaxation. Of course, this one-dimensional problem proves nothing about the more complicated two-dimensional case. All it does is give some clues (underrelaxation) and some confidence. Someone with an absolute respect for mathematical proof would not proceed before the validity of the iteration scheme was established beyond a shadow of a doubt. A strict obedience to this rule would greatly hamper the industrial mathematician and prevent him/her from making any great strides forward. In this respect we may recall Heaviside's slightly facetious remark about refusing one's dinner when one does not know the process of digestion.

Let us be clear about what was said above. There is no need for any antagonism between pure mathematics and applied/industrial mathematics. When the rules of the one are strictly enforced on the other, then trouble arises. What one should recognize is that the goals of the two disciplines are entirely different. Pure mathematicians work in a much neater and well-defined world. Their problems are defined in such a way that proofs can (almost) always be given, even if only by the best among them. In other fields, and industrial mathematics is one of them, one must always reckon with a certain measure of uncertainty. Economists are quite happy when their models offer a slightly-better-than-fifty-percent chance of predicting future occurrences correctly. To be sure, the average industrial mathematician is far more finicky than that, but more often than not he cannot meet the standards set by pure mathematicians. But then, had Aldous Huxley obeyed the grammarians' rule that every sentence requires a verb, then the first 'sentence' of 'Brave New World' would not have had the appeal it now has.

Derivation of the solution

As a working hypothesis let us assume that we cannot find a proof for the correctness of the proposed iteration scheme for the solution of our nonlinear boundary-value problem. Nevertheless, we shall proceed as dyed-in-the-wool industrial mathematicians. The problem we have to solve has a cylindrical geometry. Although the mathematics of the problems we are going to solve are comparatively trivial in principle, the final formulas look fairly complicated. Since the purpose of our presentation is to give the reader an idea of how an industrial mathematician may work, we shall proceed by studying the corresponding plane case. The final formulae are then slightly less daunting. We still consider the interval $0 \leq R \leq 1$, but now $R=0$ is a plane of symmetry, instead of an axis of symmetry.

The field equation now reads

$$\frac{\partial}{\partial R}\Lambda\frac{\partial\theta}{\partial R} + \frac{\partial}{\partial Z}\Lambda\frac{\partial\theta}{\partial Z} = 0 \ , \quad \left(\begin{array}{c} 0 \leq R \leq 1 \\ -\infty < Z < \infty \end{array}\right) \ . \tag{34}$$

For the sake of completeness we must introduce a boundary condition in the plane $R = 0$. Since this is a plane of symmetry, we have

$$\frac{\partial\theta}{\partial R} = 0 \quad \text{on} \quad R = 0 \ . \tag{35}$$

The problem in the 'crucible proper' is now defined by Eq. (34) with $\Lambda = 1$ and by Eq. (20) and

$$\frac{\partial\theta}{\partial R} = g(Z) \quad \text{on} \quad R = R_0 + 0 \ . \tag{36}$$

The function $g(Z)$ is unknown initially, but we shall try to approximate it more and more accurately as the iteration process proceeds. In view of what was said before, we put $g \equiv 0$ initially.

The solution of the problem defined above can be written as follows

$$\theta = \theta_1 + \theta_2 \ , \tag{37}$$

where

$$\theta_1 = \frac{\theta_M}{1 - R_0}\sum_{j=0}^{\infty}(-1)^j\cos\left\{\pi(j+\frac{1}{2})\frac{R - R_0}{1 - R_0}\right\}\int_{-\infty}^{\infty}e^{-\pi\frac{j+\frac{1}{2}}{1-R_0}|Z-\varsigma|}f(R_l\varsigma)d\varsigma \ , \tag{38}$$

and

$$\theta_2 = -\frac{1}{\pi}\sum_{j=0}^{\infty}\frac{(-1)^j}{j+\frac{1}{2}}\sin\left\{\pi(j+\frac{1}{2})\frac{1 - R}{1 - R_0}\right\}\int_{-\infty}^{\infty}e^{-\pi\frac{j+\frac{1}{2}}{1-R_0}|Z-\varsigma|}g(\varsigma)d\varsigma \ . \tag{39}$$

Since the function $f(R_l\varsigma)$ is given right from the start, θ_1 is given once and for all. On the other hand, $g(Z)$ changes with every iteration step and therefore θ_2 along with it.

Eq. (37), with (38) and (39), defines a temperature in the plane $R = R_0$

$$\theta(R_0, Z) = \theta_1(R_0, Z) + \theta_2(R_0, Z) \stackrel{def}{=} H(Z) \ . \tag{40}$$

This results in a boundary-value problem in the interval $0 \leq R \leq R_0$ given by Eq. (34) with Λ given by Eq. (27) and Eqs. (35) and (40). Since Λ is now a function of θ, the problem is nonlinear. Nevertheless, we are still able to solve this problem analytically. Introducing a new dependent variable, using the so-called Kirchhoff transformation

$$Q = \int_1^\theta\Lambda(\theta)d\theta = \begin{array}{ll} \theta - 1 & \text{for} \ \theta > 1 \\ \Lambda_c(\theta - 1) & \text{for} \ \theta < 1 \ , \end{array} \tag{41}$$

we may transform Eq. (34) into

$$\frac{\partial^2 Q}{\partial R^2} + \frac{\partial^2 Q}{\partial Z^2} = 0 \ . \tag{42}$$

Referring to Eq. (32), we can show that (41) not only guarantees the continuity of Q across the solid-liquid interface, but also that of the gradient $\partial Q/\partial n$. Then it follows from (42) that all higher derivatives are continuous, so that Q is analytic near the s-l interface. The boundary condition (40) is now written as

$$Q = Q_w(Z) = \begin{matrix} H(Z) - 1 & \text{for} & H > 1 \\ \Lambda_c \{H(Z) - 1\} & \text{for} & H < 1 \end{matrix} . \tag{43}$$

Further we have

$$\frac{\partial Q}{\partial Z} = 0 \quad \text{on} \quad R = 0 . \tag{44}$$

The solution to (42)-(44) can be written as follows

$$Q = \frac{1}{R_0} \sum_{j=0}^{\infty} (-1)^j \cos\left\{\pi(j + \frac{1}{2})\frac{R}{R_0}\right\} \int_{-\infty}^{\infty} e^{-\pi(j+\frac{1}{2})\frac{|Z-\varsigma|}{R_0}} Q_w(\varsigma) d\varsigma . \tag{45}$$

Using (45) together with (21), we have

$$g(Z) = \frac{\partial \theta}{\partial R}(R_0 + 0, Z) = \Lambda_b \frac{\partial \theta}{\partial R}(R_0 - 0, Z) = \Lambda_m \frac{\partial Q}{\partial R}|_{R=R_0} , \tag{46}$$

where we have used (21), (27), (29), (35) and (40). Further

$$\Lambda_m = \lambda_m/\lambda_a . \tag{47}$$

In general, since $\lambda_a \ll \lambda_m$, Λ_m is a small parameter. Since Q is an order-one function, the function $g(Z)$ is seen to be much smaller than unity, showing that our setting it equal to zero before the first iteration step was justified.

Many mathematicians would now consider their job done, of course apart from the yet unsettled matter concerning the convergence proof. However, since we are dealing here with industrial mathematics, we must still consider the question of how to get numerical values from the formulas we derived a moment ago. Trying to do this, we shall soon notice that the series of Eqs. (38), (39) and (45) are extremely impractical. Many tens of thousands or even millions of terms are needed to achieve answers that have some measure of accuracy. Adding to this that these slowly-converging series are to be used in an iteration scheme, we are led to believe that the method proposed is unsuitable. Of course, the more gullible among the industrial mathematicians may still decide to go on and request permission to use the CRAY for a whole weekend. A better thing to do would be to use a fraction of that weekend for some thinking. Indeed, by applying a trick to those dreadful series we may make them at once very suitable for numerical computation. Let us consider the series of Eq. (45). Using a simple transformation we may rewrite the integral appearing in (45) as follows

$$I = \frac{R_0}{\pi(j + \frac{1}{2})} \int_{-\infty}^{\infty} e^{-|w|} Q_w\left(Z + \frac{R_0}{\pi(j + \frac{1}{2})}w\right) dw. \tag{48}$$

We are interested in the behaviour of I for large values of j. Let us suppose that Q_w is bounded on the entire interval $-\infty < Z < \infty$ and analytic almost everywhere. Should

Q_w not be analytic in a finite number of points, then a piece-wise approach will yield similar results.

Let us now select j_0 and $w_0 \gg 1$ such that

$$\frac{R_0 w_0}{\pi(j_0 + \frac{1}{2})} \ll 1 \ . \tag{49}$$

We now write (48) as follows ($j \geq j_0$):

$$\frac{\pi(j + \frac{1}{2})}{R_0} I = \left(\int_{-w_0}^{w_0} + \int_{-\infty}^{-w_0} + \int_{w_0}^{\infty} \right) e^{-|w|} Q_w \left(Z + \frac{R_0}{\pi(j + \frac{1}{2})} w \right) dw. \tag{50}$$

The first integral can be expanded because of (49):

$$\int_{-w_0}^{w_0} e^{-|w|} \left\{ Q_w(Z) + \frac{R_0}{\pi(j + \frac{1}{2})} w Q'_w(Z) + \frac{R^2}{2\pi^2(j + \frac{1}{2})^2} w^2 Q''_w(Z) + \cdots \right\} dw \ , \tag{51}$$

whereas the absolute values of the second and the third are smaller than

$$\sup(Q_w) e^{-w_0} \ . \tag{52}$$

Choosing w_0 large enough, which is always possible for large-enough values of j_0, we may disregard (52) in the numerical computations. We can now write (45) as follows

$$R_0 Q = \left(\sum_{j=0}^{j_0} + \sum_{j=j_0+1}^{\infty} \right) (-1)^j \cos \left\{ \pi(j + \frac{1}{2}) R/R_0 \right\} I =$$

$$\sum_{j=0}^{j_0} (-1)^j \cos \left\{ \pi(j + \frac{1}{2}) R/R_0 \right\} I + \sum_{j=j_0+1}^{\infty} (-1)^j \cos \left\{ \pi(j + \frac{1}{2}) R/R_0 \right\}$$

$$\frac{R_0}{\pi(j + \frac{1}{2})} \left\{ 2 Q_w(Z) + O(j + \frac{1}{2})^{-2} \right\} =$$

$$\sum_{j=0}^{j_0} (-1)^j \cos \left\{ \pi(j + \frac{1}{2}) R/R_0 \right\} \left\{ I - \frac{2R_0}{\pi(j + \frac{1}{2})} Q_w(Z) \right\}$$

$$Q_w(Z) \sum_{j=0}^{\infty} (-1)^j \cos \left\{ \pi(j + \frac{1}{2}) R/R_0 \right\} \frac{2R_0}{\pi(j + \frac{1}{2})}$$

$$+ O \left(\sum_{j=j_0+1}^{\infty} (-1)^j \frac{\cos \left\{ \pi(j + \frac{1}{2}) R/R_0 \right\}}{(j + \frac{1}{2})^3} \right) \ . \tag{53}$$

If $Q_w(Z) \equiv 1$, then the solution in the complete domain, given by $0 < R < R_0$ and $-\infty < Z < \infty$, is identical to unity. Applying this to Eq. (45) we obtain

$$\sum_{j=0}^{\infty} (-1)^j \frac{\cos \left\{ \pi(j + \frac{1}{2}) R/R_0 \right\}}{j + \frac{1}{2}} = \frac{\pi}{2} \ , \quad (0 < R < R_0) \ . \tag{54}$$

Of course, this trivial result appears in any book on trigonometric series. However, the trick we used to arrive at (54) can also be used in more complicated cases for which these reference books do not have a ready answer. The infinite series which appears as the argument of the order symbol in Eq. (53) can be approximated as follows

$$\left| \sum_{j=j_0+1}^{\infty} (-1)^j \frac{\cos\left\{\pi(j+\frac{1}{2})R/R_0\right\}}{(j+\frac{1}{2})^3} \right| < \sum_{j=j_0+1}^{\infty} \frac{1}{(j+\frac{1}{2})^3}$$

$$< \int_{j_0}^{\infty} \frac{d\sigma}{(\sigma+\frac{1}{2})^3} = \frac{1}{2(j_0+\frac{1}{2})^2} \ . \tag{55}$$

As a result we find the following representation for the expression of Eq. (45):

$$Q = Q_w + \sum_{j+0}^{j_0} (-1)^j \left\{ \frac{I}{R_0} - \frac{2}{\pi(j+\frac{1}{2})} Q_w(Z) \right\} \cos\left\{ \pi(j+\frac{1}{2})\frac{R}{R_0} \right\}$$

$$+ O(j_0^{-2}) \ . \tag{56}$$

Whereas, using (45), we may need many tens of thousands of terms to calculate four significant digits of Q, we need only about one hundred when we use (56). Of course, when a more accurate result is required, we can always extend the analysis that led to Eq. (56) by including higher derivatives of Q_w.

Results

This exercise in industrial mathematics will not be complete before results are presented which can be understood by crystal growers. Clearly, a bunch of formulas with a recipe on how to get numbers from them will not do. We shall have to do the numerical programming ourselves, carrying out the iteration process so as to get an insight into the shape of the solid-liquid interface. We shall not describe at length here what these numerics look like. They are fairly straightforward and not particularly exciting. It should be pointed out that one must be careful in dealing with the discontinuity in the longitudinal derivative of Q_w which is caused by the transformation of Eq. (43). During the iteration process the discontinuity will change position, so that the discretisation of the Z-axis will have to be carried out adaptively.

Figs. 5a, b, c give an idea of the results the present system may yield. These graphs show the shape of the solid-liquid interface for three different wall temperatures. Denoting the dimensionless outer wall temperature by θ_w, we have

$$\theta_w(Z) = \theta_M \exp\{-R_l^2 Z^2\} \ . \tag{57}$$

Choosing $R_\varrho = 2^{-\frac{1}{2}}, 1, 2^{\frac{1}{2}}$ respectively, we can simulate three different lengths of the heat source (rf coil). As we said before, the length of the heating coil is a matter that sometimes leads to emotional debates among crystal growers. The figures show that narrow heat sources (Fig. 5c) require higher maximum temperatures (θ_M) to yield relatively

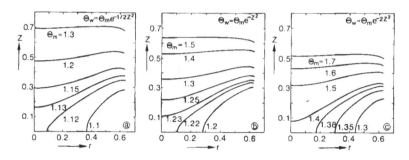

Figure 5: Solid-liquid interfaces for three different wall-temperature profiles.

flat s-l interfaces. Fig. 5c shows that at $\theta_M = 1.3$, which means a maximum temperature that is 30% above the melting temperature, only part of the raw materials (powder) are melted. Clearly, suitable crystals cannot be grown under such conditions. A wider heat source (Fig. 5a) yields a more-or-less flat interface at $\theta_M = 1.3$. This makes a strong case for those crystal growers who favour extended mildly varying heat sources.

It is also clear from Figs. 5a-c that the s-l interface undergoes strong displacements in certain narrow temperature ranges. For instance, Fig. 5a shows that, when θ_M changes from 1.1 to 1.15, the interface experiences a large displacement. This will be a cause of instabilities, since there are always temperature fluctuations. Although these fluctuations seem to be less pronounced when the s-l interface is relatively flat, this effect is still something an experimenter has to think about.

This discussion shows that it is not always necessary to describe an experimental situation exactly if one wants to reveal and investigate some of its more important aspects. In such cases we are concerned with obtaining a qualitative insight which may persuade a crystal grower to change his set-up. When, on the basis of intuition, he has already taken the right precautions, then the confirmation afforded by a mathematical model will increase his trust in them. In both cases his judgement about the mathematical approach will be positive.

We remarked earlier that a flat solid-liquid interface is the ideal situation for a crystal grower. Figs. 5a-c suggest that achieving this is only partly possible. The interfaces always seem curved when they approach the inner wall of the crucible. Of course, one could say that these graphs represent only a handful of special cases, and that it may still be possible to achieve an absolutely flat interface by choosing the correct wall temperature. However, a moment's thought will suffice to convince us that this ideal situation can never be achieved in a Bridgman-Stockbarger set-up. Indeed, let us suppose for argument's sake that the s-l interface were flat. In that case the temperature distribution within the charge region, but at the inner crucible wall, will experience a sudden change in its slope which can be attributed to the thermal conductivity having different values in the melt and the solid. Approaching the inner crucible wall from the other side, i.e. at $R = R_0 + 0$, we find a temperature profile with a continuously varying slope, since the thermal conductivity λ_c is a constant. This creates a conflict, since the temperature distributions at $R = R_0 - 0$

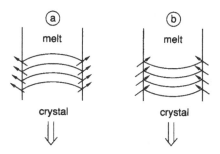

Figure 6: Propagation of defects.

and $R = R_0 + 0$ are the same. Carrying our argument still further we find that the behaviour of the s-l interface is singular in the vicinity of the point where it meets the inner wall of the crucible. We refer to the literature for a discussion of this singularity.

The unavoidable strong curvature of the s-l interface near the crucible wall may be a source of thermal tension which in turn may induce defects in the crystal. Depending upon the shape of the interface, such defects may or may not propagate into the crystal (Fig. 6). This is another reason why crystal growers take a strong interest in the shape of these interfaces.

What has this case study taught us?

1. That a problem, or rather a situation, which at first seemed hopelessly complicated, may be transformed into something that we can deal with. What one needs is the courage to deviate from the original problem definition and to strive for less ambitious goals;

2. That it is useful to make a list of all the relevant parameters and variables and to keep track of their dimensions;

3. That one will profit a great deal from making a problem dimensionless. More often than not this procedure shows one the way to simplify a problem further;

4. That one must have the courage, some would say lack of conscience, to proceed with one's actions, even if some of these have not yet received the stamp of approval afforded by mathematical proof;

5. That an investigation in industrial mathematics is not complete once the final formulas have been given. One must also produce the necessary numerical values and present these in a form which can be appreciated by non-mathematicians. In this

respect graphs, tables and a discussion and interpretation presented in ordinary language are very useful;

6. That a thorough investigation of an industrial-mathematics nature raises more new questions than it answers. Strictly speaking, since reality can never be modelled to the full, any attempt at modelling it completely will be a never-ending process.

Appendix: *A one-dimensional example illustrating the iteration procedure.*

Let us consider one-dimensional conduction between two planes denoted by $Y = 0$ and $Y = 1$ respectively. Between these planes there is another plane at $Y = \eta$ which separates two regions of different thermal conductivities. The region $0 < Y < \eta$ has a thermal conductivity equal to Λ, that in the interval $\eta < Y < 1$ has a thermal conductivity equal to unity. We shall assume that $\Lambda \ll 1$. The boundary conditions for the temperature θ are

$$\theta = 0 \quad \text{at} \quad Y = 0, \quad \theta = 1 \text{ at } Y = 1. \tag{A1}$$

Continuity of temperature and heat flux at the interface is expressed as follows

$$\theta(\eta - 0) = \theta(\eta + 0), \tag{A2}$$

$$\Lambda \frac{\partial \theta}{\partial y}(\eta - 0) = \frac{\partial \theta}{\partial y}(\eta + 0) = g, \tag{A3}$$

where g is the dimensionless heat flux that we shall need in the iteration process. The solution to this trivial problem is

$$\theta = \begin{cases} \frac{Y}{\eta + \Lambda(1-\eta)} & \text{for } 0 < Y < \eta \\[2mm] \frac{\eta + \Lambda(Y - \eta)}{\eta + \Lambda(1-\eta)} & \text{for } \eta < Y < 1. \end{cases} \tag{A4}$$

We shall now try to obtain this solution by iterative means. Expressed in the still unknown heat flux g (see A3), we find a temperature

$$\theta = 1 + g(Y - 1) \text{ for } \eta < Y < 1. \tag{A5}$$

Using (A1) and (A2) we also find

$$\theta = \{1 + g(\eta - 1)\}\frac{Y}{\eta} \quad \text{for} \quad 0 < Y < \eta. \tag{A6}$$

Now supposing that $\Lambda \ll 1$, we put $g = g_0 = 0$ in the first iteration step. Using (A6) and then (A3) we find a new value for g which we denote by g_1. This process can be continued *ad infinitum*. The general recurrence relation is

$$g_{i+1} = \{1 + g_i(\eta - 1)\}\frac{\Lambda}{\eta} \quad \text{with } g_0 = 0. \tag{A7}$$

The general solution of this recurrence is

$$g_i = \frac{\Lambda}{\eta + \Lambda(1 - \eta)} \left\{ 1 - \left(\Lambda \frac{\eta - 1}{\eta} \right) \right\} . \tag{A8}$$

Clearly, this leads to a limit value only if

$$\Lambda < \frac{\eta}{1 - \eta} , \tag{A9}$$

showing that the iteration process converges as long as Λ is small enough.

Using underrelaxation we can induce convergence also for those cases where Λ does not satisfy (A9). Instead of writing (A7), we use

$$g_{i+1} = \omega\{1 + g_i(\eta - 1)\}\frac{\Lambda}{\eta} + (1 - \omega)g_i , \tag{A10}$$

again with $g_0 = 0$. The factor ω is called the underrelaxation parameter. It has to satisfy the condition $0 < \omega < 1$. The iteration process of (A10) converges if

$$-1 < \omega\frac{\Lambda}{\eta}(\eta - 1) + 1 - \omega < 1 . \tag{A11}$$

The optimal result is obtained when

$$\omega = \frac{\eta}{\eta + \Lambda(1 - \eta)} , \tag{A12}$$

showing that strong underrelaxation (ω small) must be applied when η is very small or when Λ is very large. When Λ is small, then $\omega \sim 1$, indicating, as expected, that there is no need for underrelaxation.

Suggestions for further reading

On temperature calculations for Bridgman-Stockbarger configurations:

- R.J. Naumann & S.L. Lehoczky, J. Crystal Growth **61**, 707-710 (1983)

- T. Jasinski & A.F. Witt, J. Crystal Growth **67**, 173-184 (1984)

- T. Jasinski & A.F. Witt, J. Crystal Growth **71**, 295-304 (1985)

On the wall singularity:

- H.K. Kuiken, SIAM J. Appl. Math. **48**, 921-924 (1988)

On the application of the iterative method based on partial analytic solutions to a more complicated 3-d problem:

- M.J.J. Theunissen, R.P.M.L.C. van de Nieuwenhoff & H.K. Kuiken, J. Appl. Phys. **68**, 806-813 (1990).

Thermal behaviour of a high-pressure gas-discharge lamp

A high-pressure gas-discharge lamp consists of a transparent vessel which may be cylindrical, spherical or otherwise, containing a gas which is at a high pressure at the operating temperature. Within this vessel there are two electrodes between which an electric field can be set up. Electrons leave one of the electrodes, are accelerated by the electric field and move towards the other electrode. Since the gas is at a high pressure, the mean free path is small and collisions between electrons and gas molecules are frequent. These collisions can be of various kinds. Most collisions result in the acceleration of the gas atoms, thereby increasing their temperature. Some collisions result in the excitation of the valence electrons. These electrons are likely to return to lower energy levels. This process is accompanied by the emission of photons. For some gases, such as mercury or sodium vapour, these photons are in the visible range or they can easily be transformed into visible photons.

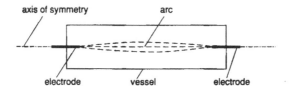

Figure 1: Geometrical model of gas discharge

We remarked already that most electron-gas collisions result in the acceleration of the gas atoms. In each instance the energy increment will be small, owing to the large difference between the masses of the two colliding particles. Since the electrons soon regain their previous energy levels, due to their low mass, and since collisions are extremely frequent, the temperature will be very high in a high-pressure gas-discharge lamp. In the centre of the arc, where most of the energy is dissipated, the temperature may reach values exceeding 5000 K. On the other hand, the temperature of the vessel wall may not be too high, typically 1500 K at the most. If one realises that the inside of these vessels is often not wider than a few millimetres, one can appreciate that temperature control within high-pressure gas-discharge lamps is of the utmost importance.

A simple thermal conduction model

We shall consider a tube with inner radius a (Fig. 1). The tube is slender, i.e. it is much longer than a. Between two electrodes, which are situated at either end of the tube, there exists an electric field E which is independent of the radial coordinate r. Denoting the axial coordinate by x, we assume that E is also independent of x, except, of course, in the immediate neighbourhood of the electrodes. In our model we shall restrict our attention to the relatively long region which excludes the two end regions of the tube. In this region E is truly uniform.

Owing to the presence of the electric field, a stream of electrons will flow from the negatively charged electrode to the anode. This is the same as saying that there is an electric current density

$$i = \sigma(t)E , \tag{1}$$

where $\sigma(t)$ is the conductivity which is a function of the temperature t. The mathematical model we wish to formulate will attempt to describe the steady state at the operating temperature of the lamp. When convection is excluded, this temperature will be a function of r only.

Integrating (1) across the tube cross section, we obtain the total current

$$I = 2\pi E \int_0^a r\sigma(t)dr , \tag{2}$$

which can easily be measured, since the lamp is part of a circuit. The energy dissipation per unit volume is defined by

$$q = i E . \tag{3}$$

Using (1) and (2) we may write this as

$$q = \frac{I^2}{4\pi^2} \frac{\sigma(t)}{(\int_0^a r\sigma(t)dr)^2} . \tag{4}$$

In a simple model, where conduction is the only means of thermal transport, the temperature is governed by

$$\nabla \lambda(t) \nabla t + q(t) = 0 , \tag{5}$$

where $\lambda(t)$ is the temperature-dependent thermal conductivity. In this model radiative losses have been neglected. This is a reasonable assumption for radiation phenomena appertaining to the visible range, since only a small portion of the energy transfer is related to this part of the spectrum. On the other hand, there is a considerable amount of radiative transfer between the various parts of the gas itself. Certain photons which are emitted in the central part of the arc are absorbed after they have travelled only a very short distance. The absorbed energy may be reemitted as photons, or it may lead to a local heating effect. This radiation is called optically thick, as opposed to the thin radiation which leaves the arc and the cylindrical region without being absorbed. The thermal effect of optically thick radiation is often modelled by an increase of the thermal conductivity $\lambda(t)$ which is written as

$$\lambda(t) = \lambda_{\text{cond}} + \lambda_{\text{rad}} . \tag{6}$$

Another effect neglected in (5) is convection.

The ultimate purpose of our model will be to determine the temperature drop between the axis of the cylinder and the wall. Therefore we shall prescribe the temperature on the wall of the cylinder, which can often be measured easily. Therefore

$$t = t_w \quad \text{at} \quad r = a . \tag{7}$$

On this axis we have

$$\frac{dt}{dr} = 0 \quad \text{at} \quad r = 0 , \tag{8}$$

where we used 'straight' d's since the temperature is a function of r only. To complete the model, we must prescribe $\lambda(t)$ and $\sigma(t)$ as functions of temperature. For many practical purposes we can write

$$\lambda(t) = \lambda_r (t/t_r)^{\frac{3}{4}} , \tag{9}$$

$$\sigma(t) = \gamma t^{\frac{3}{4}} \exp(-t_i/t) , \tag{10}$$

where t_r is some reference temperature which is still to be defined and t_i is the so-called ionisation temperature

$$t_i = eV_i/2k . \tag{11}$$

In (11) e is the charge of an electron, V_i is the ionisation potential and k is Bolzmann's constant. Further, γ is a constant which is inversely proportional to the square root of the gas pressure and the collision cross section of the mercury atoms.

Dimensionless formulation

The reader has probably noticed already that our model has led to a non-linear ordinary differential equation for the single unknown t. Some researchers would now opt for a numerical routine, plugging in the parameter values as they present themselves. Whereas this approach will give a quick answer for a particular practical situation, it is not very helpful in giving the researcher new insight. A mathematician may easily overlook the fact that quite often engineers or industrial designers are not interested in precise numerical solutions of models which are in themselves only gross approximations of reality, but rather in the trends these models may reveal.

A powerful means of bringing forth the true essence of a problem is dimensional analysis which is nothing but rendering the problem dimensionless in a judicious manner. In our simple problem there are only two variables that have to be made dimensionless, viz. r and t. It would seem obvious that r shall be made dimensionless with the radius of the tube, which is a. It is more difficult to choose a proper characteristic temperature. The problem definition itself contains two temperatures, namely t_w and t_i. We are still free to choose the temperature t_r occurring in eq. (9) and we can replace it by any value by redefining λ_r. The ionisation temperature t_i is of the order of many tens of thousands of degrees Kelvin, which is much higher than the expected maximum temperature in the arc. If t_i is chosen as the reference temperature, the dimensionless temperature is very much smaller than unity, which is not attractive. The same is true for t_w in an inverse sense. Choosing t_w for our characteristic temperature, we obtain a dimensionless temperature which is very much larger than unity in the important central part of the arc, which is not very attractive either. We, therefore, make the unusual step of making t dimensionless with the, as yet unknown, temperature on the axis of the cylinder, which we denote by t_r. Therefore, the parameter λ_r occurring in (9) is now the thermal conductivity right in the centre of the arc.

Having said this, we introduce

$$R = r/a \ , \quad T = t/t_r \ . \tag{12}$$

It is easily seen that the problem is now defined by the following equation and boundary conditions

$$\frac{1}{R}\frac{d}{dR}RT^{\frac{3}{4}}\frac{dT}{dR} + HT^{\frac{3}{4}}\exp\left\{T_i(1 - \frac{1}{T})\right\} = 0 \ , \quad (0 \le R \le 1) \ , \tag{13}$$

$$T = 1 \ , \quad \frac{dT}{dR} = 0 \quad \text{at} \quad R = 0 \ , \tag{14}$$

$$T = \beta T_i \quad \text{at} \quad R = 1 \ , \tag{15}$$

where

$$T_i = t_i/t_r \ , \tag{16}$$

$$H = \alpha f(T_i) \ , \tag{17}$$

$$f(T_i) = T_i^{\frac{5}{2}}\, e^{\,T_i} \left(\int_0^1 RT^{\frac{3}{4}}\exp\{T_i(1 - T^{-1})\}dR\right)^{-2} \ , \tag{18}$$

$$\alpha = \frac{I^2}{4\pi^2 a^2 \gamma \lambda_i t_i^{\frac{7}{4}}} \ , \tag{19}$$

$$\beta = \frac{t_w}{t_i} \ . \tag{20}$$

Since (13) is a second-order differential equation, albeit a strongly nonlinear one, only two boundary conditions can be prescribed in general. But then, Eqs. (14) and (15) prescribe three conditions. However, there is still the unknown scaling factor t_r to contend with. As a result T_i, and therefore H, are unknown parameters to begin with. It can be seen from (17) and (19) that the factor α intervenes between H and T_i. This offers an interesting way of solving this problem. We can *give* the value of T_i and then ask ourselves which value of H is needed to satisfy the third boundary condition. In this way we are implicitly asking for the correct value of the multiplicative factor α and therefore the precise value of the current I necessary to produce the aforesaid value of T_i. This is, in a way, an inverse formulation of the problem, namely which current is needed to produce a particular axial temperature. As a bonus, this approach offers an easy way out of the tricky problem of having to solve the integral in (18) *during* the integration procedure. Now we have to evaluate it only after the correct dimensionless temperature profile has been evaluated.

Summarizing, we give T_i. Then we solve the system (13-15). This leads to

$$H = H(\beta, T_i) \ , \tag{21}$$

where β is only a passive, i.e. given parameter, and a given temperature profile. Then we put

$$\alpha = \frac{H(\beta, T_i)}{f(T_i)} \ , \tag{22}$$

defining the current parameter α as a function of the axis temperature t_r through $T_i = t_i/t_r$. Alternatively, we may invert the function (22), finding t_r as a function of the current parameter α.

A model within the model

The dimensionless model given by (13)-(15) is already far more attractive than the original dimensional one. Indeed, there are only two parameters left, viz. T_i and β, which characterize the complete problem. Again, some people would now decide to proceed by applying numerical methods, since, obviously, the problem cannot be solved analytically. However, before embarking upon such a course, it is wise for us to consider the numerical ranges of the parameters T_i and β first. Of course, I refer to the practical values these parameters can have, not to those that are mathematically possible. In principle, those parameters can assume any values, as long as these are positive, but in every practical situation T_i will be much larger than unity and β will be much smaller than unity. Therefore, the natural way to approach this problem is to apply *asymptotic* methods.

When T_i is large, the exponential function in (13) varies very rapidly. The argument is equal to zero at $R = 0$, since $T(0) = 1$. However, when T decreases, the argument increases rapidly. According to (15), the temperature at the other end of the interval is equal to βT_i which is very much smaller than unity. Indeed, $\beta T_i = t_w/t_r$, which for practical situations has a value of 0.1-0.2. With $T_i \gg 1$, the argument of the exponential function will attain values which may be of the order of 50-100 near the tube wall. Clearly, this analysis shows that the heating term only plays a role in a very narrow region around the axis. Elsewhere it can be put equal to zero, leaving an equation in which only conduction is represented.

To obtain insight into this asymptotic behaviour we shall study a simpler problem. We shall devise, as it were, a model within the model. Equation (13) is complicated somewhat by the rotational symmetry, and by the temperature-dependent thermal conductivity. These two aspects of the problem do not seem to be essential for the two-layer phenonemon described above. We shall therefore consider the simpler equation

$$\frac{d^2T}{dY^2} + H \exp\left\{T_i\left(1 - \frac{1}{T}\right)\right\} = 0 \ , \quad (0 \le Y \le 1) \ , \tag{23}$$

$$T = 1 \ , \quad \frac{dT}{dY} = 0 \quad \text{at} \quad Y = 0 \ , \tag{24}$$

$$T = 0 \quad \text{at} \quad Y = 1 \ . \tag{25}$$

which expresses a plane geometry. Moreover we put $\beta = 0$.

Asymptotics for $T_i \rightarrow \infty$

If we consider any region where $T < \tilde{T} < 1$, where $\tilde{T}$ is a fixed value, and let $T_i \rightarrow \infty$, then the exponential term of (23) is exponentially small in that region. As a consequence of this, the temperature profile in that region is governed by $d^2T/dY^2 = 0$, so that

$$T = A(1 - Y) = T_{outer} \ , \tag{26}$$

where A is an, as yet, undetermined constant. Since $T = 0$ clearly belongs to the aforesaid region, Eq. (26) satisfies Eq. (25).

To describe what happens in the origin close to $Y = 0$ where $T = 1$, we can allow values of T for which the argument, when $T_i \rightarrow \infty$, remains finite. This is achieved by introducing

$$T = 1 - \frac{Q}{T_i} = T_{inner} \ , \tag{27}$$

where Q is a function which remains finite when $T_i \rightarrow \infty$. Substituting (27) in (23) and retaining terms of the highest order only, we obtain

$$\frac{d^2Q}{dY^2} = \alpha e^{-Q} \ . \tag{28}$$

Because of (24), Q must satisfy the conditions

$$Q = 0 \ , \quad \frac{dQ}{dY} = 0 \quad \text{at} \quad Y = 0 \ . \tag{29}$$

Further,

$$\alpha = HT_i \ . \tag{30}$$

The solution to (28) which satisfies (29) is

$$Q = 2\ln\left\{\cosh\left(Y\sqrt{\frac{\alpha}{2}}\right)\right\}. \tag{31}$$

We still have to determine the constants α and A. This is done by means of a so-called matching technique. The outer solution (26) and the inner solution (27) are supposed to have a region of common validity. This is expressed as follows

$$\begin{array}{cc} \text{behaviour } T_{inner} & = & \text{behaviour } T_{outer} \\ Y\sqrt{\alpha/2} \gg 1 & & Y \ll 1 \end{array} \tag{32}$$

in the limit $T_i \rightarrow \infty$. It is understood that *fixed* values $Y \ll 1$ and $Y\sqrt{\frac{\alpha}{2}} \gg 1$ are used before the limit $T_i \rightarrow \infty$ is taken. When we apply the matching rule to (26) and (27), we find

$$A = 1 \ . \tag{33}$$

Since both the inner and the outer differential equation for T are of the second order, we must also demand the matching of the derivative, i.e.

$$-1 = \lim_{T_i \rightarrow \infty}\left\{-\frac{1}{T_i}\frac{dQ}{dY}\Big|_{Y\sqrt{\frac{\alpha}{2}} \gg 1}\right\} \tag{34}$$

or

$$-1 = -\lim_{T_i \to \infty} \frac{\sqrt{2\alpha}}{T_i} \tanh\left(Y\sqrt{\frac{\alpha}{2}}\right) . \tag{35}$$

Clearly, the tanh function is equal to unity, apart from exponentially small terms. Therefore

$$\alpha = \frac{T_i^2}{2} , \tag{36}$$

or, using (30),

$$H = \frac{1}{2}T_i . \tag{37}$$

It is clear from (31) and (36) that the inner region is described properly by the scaled variable

$$Z = T_i Y . \tag{38}$$

The properly scaled inner solution is then given as

$$T_{\text{inner}} = 1 - \frac{2}{T_i} \ln\left(\cosh(\frac{Z}{2})\right). \tag{39}$$

A composite expansion which is a uniformly valid approximation to the full solution in the complete closed interval $0 \leq Y \leq 1$ is obtained from the rule

$$T_{\text{comp}} = T_{\text{inner}} + T_{\text{outer}} - \text{common part.} \tag{40}$$

The common part is the part which was obtained during the matching procedure. Since we matched both terms of $T_{\text{outer}} = 1 - Y$, we find

$$T_{\text{comp}} = 1 - \frac{2}{T_i} \ln \cosh(\frac{Z}{2}) , \tag{41}$$

which is identical with the inner solution found so far. To see what this composite solution yields for fixed values of Y which are anywhere in the interval $0 \leq Y \leq 1$ we use (38) and conclude that Z is large for those values of Y. Therefore

$$T_{\text{comp}} \sim 1 - \frac{2}{T_i} \left\{ \frac{Z}{2} - \ln 2 + \text{EST} \right\} , \tag{42}$$

where EST stands for exponentially small terms. Using (38), and disregarding EST, we can write (42) as follows

$$T_{\text{comp}} \sim 1 - Y + \frac{2 \ln 2}{T_i} , \tag{43}$$

showing that the composite expansion is indeed correct up to terms of $O(T_i^{-1})$. Higher-order asymptotics are needed to obtain a more accurate solution.

Higher-order asymptotics

It should be clear from the foregoing section that the proper transformation defining the inner solution is

$$T(Y) = T_{inner} = 1 - \frac{1}{T_i} \sum_{n=0}^{N} \frac{Q_n(Z)}{T_i^n} , \tag{44}$$

$$Z = Y T_i , \tag{45}$$

$$H = T_i \sum_{n=0}^{N} \frac{1}{T_i^n} H_n , \tag{46}$$

where

$$Q_0(Z) = 2 \ln \left\{ \cosh(\tfrac{1}{2} Z) \right\} , \tag{47}$$

$$H_0 = \frac{1}{2} . \tag{48}$$

Substituting (44)-(46) in (23)-(24), we find that the next perturbation term is governed by

$$\frac{d^2 Q_1}{dZ^2} + \frac{1}{2} \cosh^{-2}(\tfrac{1}{2} Z) Q_1 = \left(H_1 - 4 \ln^2 \cosh(\tfrac{1}{2} Z) \right) \cosh^{-2}(\tfrac{1}{2} Z) , \tag{49}$$

$$Q_1(0) = 0 ; \quad \frac{dQ_1}{dZ}(0) = 0 . \tag{50}$$

After a lengthy calculation the solution of (49)-(50) can be shown to be

$$\begin{aligned}
Q_1 = & \left(-Z^2 + (H_1 + 1)Z + 4Z \ln \cosh \tfrac{1}{2} Z \right. \\
& \left. + \; 8 \int_0^{\frac{Z}{2}} u(1 - \tanh u) du \right) \tanh \tfrac{1}{2} Z \\
& - \; 4 \ln^2 \cosh \tfrac{1}{2} Z - 4 \ln \cosh \tfrac{1}{2} Z ,
\end{aligned} \tag{51}$$

From this we find

$$\begin{aligned}
\text{behaviour} \quad Q_1 = & \; (H_1 - 1)Z \; + \; 4(1 - \ln 2) \ln 2 + \tfrac{\pi^2}{3} \\
z \gg 1 \quad & \\
& + \; \text{EST} ,
\end{aligned} \tag{52}$$

where we have used the identity

$$\int_0^{\infty} u(1 - \tanh u) du = \frac{\pi^2}{24} . \tag{53}$$

Using the earlier result for Q_0 we can now write down

$$\begin{aligned}
\text{behaviour} \quad T_{inner} = & \; 1 - \tfrac{2}{T_i} \left(\tfrac{1}{2} Z - \ln 2 \right) \\
z \gg 1 \quad & \\
T_i \to \infty \quad & \\
& - \; \tfrac{1}{T_i^2} \left\{ (H_1 - 1)Z + 4(1 - \ln 2) \ln 2 + \tfrac{\pi^2}{3} \right\} + O\left(\tfrac{1}{T_i^3} \right) \\
= & \; 1 - Y + \tfrac{1}{T_i} \left\{ (1 - H_1)Y + 2 \ln 2 \right\} + O\left(\tfrac{1}{T_i^2} \right) .
\end{aligned} \tag{54}$$

The outer solution is characterized by the exponential smallness of the second term on the left of equation (23). Therefore, if we write

$$T = T_{\text{outer}} = 1 - Y + \frac{1}{T_{\text{i}}}T_1 + O\left(\frac{1}{T_{\text{i}}^2}\right) ,$$ (55)

we find on account of Eq. (25)

$$T_1 = A_1(1 - Y) ,$$ (56)

where A_1 is some constant which will be determined presently. Comparing (54) and (55) we see that the inner and the outer expansion match perfectly when

$$A_1 = 2\ln 2$$ (57)

and

$$H_1 = 1 + 2\ln 2 .$$ (58)

As before, we can write down a composite expansion by evaluating

$$T_{\text{comp}} = T_{\text{inner}} + T_{\text{outer}} .$$ (59)

In this relatively simple problem T_{outer} and the common part are the same, so that the composite expansion and the inner solution coincide.

Concluding remarks

The simple model within the model has taught us a great deal about the structure of solutions of systems such as that defined by Eqs. (23)-(25) when T_{i} is much larger than unity. We can now try and solve a similar system where the governing equation is given by

$$\frac{d}{dY}T^{\frac{3}{4}}\frac{dT}{dY} + HT^{\frac{3}{4}}\exp\left\{T_{\text{i}}\left(1 - \frac{1}{T}\right)\right\} = 0 .$$ (60)

The algebra will be slightly more difficult, but, in principle, nothing changes. The next step will be to consider the axially symmetric case which finds its expression in Eq. (13). This step is less trivial than the one which led to Eq. (60) but, no doubt, the way to solve it will be much easier to find, now that we know how to treat the simple case. A further complication will be to introduce convection and again our combined knowledge of the simpler cases will show us the way. This serves to show that a great deal of understanding and sheer enjoyment in doing some exciting mathematics is lost when we yield too early to the attractions of the computer.

Suggestions for further reading

On gas discharges and arcs:

- W. Elenbaas, *The high-pressure mercury vapour discharge*. North Holland, 1951.

- J.J. Lowke, J. Appl. Phys. **50** (1979) 147-157.

- H.K. Kuiken, Appl. Phys. Lett. **58** (1991) 1833-1835.

- H.K. Kuiken, J. Appl. Phys. **69** (1991) 2896-2903.

- H.K. Kuiken, J. Appl. Phys. **70** (1991) 5282-5291.

- R.J. Zollweg, J. Appl. Phys. **49** (1978) 1077-1091.

On singular perturbations:

- J. Kevorkian and J.D. Cole, *Perturbation methods in applied mathematics,* Springer, 1981.

- A.K. Kapila, *Asymptotic treatment of chemically reacting systems,* Pitman, 1983.

The determination of surface tension by means of the sessile-drop method

Surface tension and pressure

From everyday experience everyone is familiar with the notion of surface tension. Drops dripping from a tap, brandy creeping up against the inside of a glass, small invertebrate animals 'walking' on the surface of a pond, are all examples of phenomena that would not exist without surface tension. In the absence of surface tension a liquid body, such as a drop, would not remain a coherent whole. The phenomenon of surface tension can be attributed to the fact that molecules which are sitting at the surface have an energy level different from those in the bulk. Energy is needed to transport a molecule from the bulk to the surface, and such transport is necessary if we wish to increase the outer surface of a liquid body. Therefore, if we leave such a liquid body alone, it will always tend to make its surface as small as possible under the given circumstances. This tendency manifests itself as surface tension. If no other forces act on a liquid body, surface tension will cause it to become spherical.

Let us consider a line element dl lying within the outer surface of a liquid body (Fig. 1). Let n be a unit normal on this line element within the tangent plane. The liquid in the immediate vicinity of dl situated at the side denoted by n exerts a force $\gamma\, n\, dl$ on the liquid situated at the other side of dl. According to Newton's third law, the latter liquid will exert a force $-\gamma n dl$ on the former. The quantity γ is called the surface tension of the particular liquid under consideration. From the foregoing definition it is immediately clear that the dimension of γ is given by

$$[\gamma] = Nm^{-1} = kgs^{-2}\,, \tag{1}$$

showing that surface tension is a (pull) force per unit length.

Let us now consider a closed curve $\Gamma(\tau)$ in the outer surface of a liquid body (Fig. 1). $\Gamma(\tau)$ is a material curve which means that each of its points is affixed to a material point in the outer surface of the liquid. The argument τ denotes that the curve is given at the time τ. The part of the liquid surface enclosed by Γ has the area O. Let us suppose that the material points on the surface are displaced. An infinitesimal time dt later our curve will be denoted by $\Gamma(\tau + d\tau)$. The displacement which maps $\Gamma(\tau)$ onto $\Gamma(\tau + d\tau)$ can be described by v_n which is the normal component of a displacement-velocity vector, where n is the outward normal lying in the tangent plane. The work needed to bring about the aforementioned displacement is given by

$$dA = d\tau \int_{\Gamma(\tau)} \gamma v_n dl. \tag{2}$$

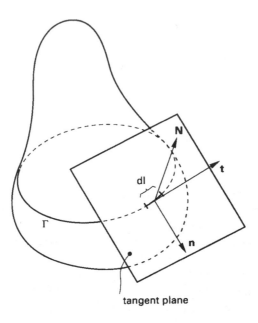

Figure 1: Geometric configuration.

The change in the area enclosed by Γ is given by

$$dO = d\tau \int_{\Gamma(\tau)} v_n dl. \qquad (3)$$

Assuming for the sake of argument that γ is a constant, we deduce from (2) and (3)

$$\frac{dA}{dO} = \gamma, \qquad (4)$$

showing that γ is equal to the work needed to induce a (positive) change in area of the outer surface.

Until now we have restricted our attention to the surface of the liquid body under consideration. If there are no fluid motions inside the body, the pressure p suffices for the description of the stress level in the interior. For the pressure p we have

$$[p] = N.m^{-2} = kg.m^{-1}.s^{-2}. \qquad (5)$$

At each point it describes how strongly the liquid is compressed. If ds be a surface element in the interior of the liquid and $\mathbf{k}$ is a unit normal, then the force exerted on the liquid at the side of $\mathbf{k}$ by the liquid at the other side is given by

$$d\mathbf{F} = pk ds, \qquad (6)$$

fully in line with the compressive character of the pressure (Fig. 2). Reversing the role of the portions of liquid defined in the previous sentence we obtain the balancing force

Figure 2: Interior surface element ds with normal **k**.

-d**F**. Now suppose that ds is not in the interior, but on the outer surface. In that case surface tension takes care of the balancing force. To describe this force balance we refer to Fig. 1. In each point of Γ, which is now assumed to be fixed, we define a set of three mutually orthogonal unit normals **n**, **t** and **N**. **N** is the outward normal to the liquid surface. Therefore, both **n** and **t** are in the tangent plane that touches the liquid surface in the particular point under consideration. We intend to write down a force balance for a part of the outer surface which is enclosed by Γ. Obviously, there can be two of them. Once we have decided which of these parts we mean, which we shall denote by Ω, **n** is the outward normal with respect to Γ. Finally, **t** is tangent to Γ, its sense being selected in such a way that **n**, **t** and **N** form a local right-handed Cartesian coordinate system.

Using the definitions of γ and p already given, we can write down the required force balance as follows:

$$\int_\Omega p\mathbf{N}ds + \int_\Gamma \gamma\mathbf{n}dl = 0. \tag{7}$$

In the analysis that follows we shall need expressions such as $div\mathbf{N}$ and $rot\mathbf{N}$ which, strictly speaking, are not defined uniquely,.since **N** has been given on Ω only. One can prove, however, that these vector operations are uniquely defined on Ω if the **N** field on Ω is embedded in an arbitrary spatial **N** field consisting solely of unit vectors.

Multiplying (7) by an arbitrary constant vector **m**, and using the fact that **n**, **t** and **N** constitute a right-handed system, we can deduce from (7)

$$\int_\Omega p\mathbf{N}.\mathbf{m}ds + \int_\Gamma \gamma(\mathbf{t} \times \mathbf{N}).\mathbf{m}dl = 0. \tag{8}$$

Using Stokes's theorem we can write the second integral as follows:

$$\int_\Gamma (\mathbf{t} \times \mathbf{N}).\mathbf{m}dl = \int_\Gamma (\mathbf{N} \times \mathbf{m}).\mathbf{t}dl = \int_\Omega \mathbf{N}.rot(\mathbf{N} \times \mathbf{m})ds , \tag{9}$$

where we have assumed γ to be a constant. When γ is not constant, Eq.(7) would not be valid, because it does not account for the fluid motions that a non-uniform γ will induce. Using a well-known rule from vector analysis and considering the fact that **m** is constant, we can write

$$rot(\mathbf{N} \times \mathbf{m}) = -\mathbf{m}div\mathbf{N} + (\mathbf{m}.grad)\mathbf{N}, \tag{10}$$

so that

$$\int_\Omega \mathbf{N}.rot(\mathbf{N} \times \mathbf{m})ds = - \int_\Omega (\mathbf{N}.\mathbf{m})div\mathbf{N}ds, \tag{11}$$

where we have used

$$\mathbf{N}.(\mathbf{m}.grad\mathbf{N}) = 0. \tag{12}$$

Eq. (12) can be proved when we realize that $\mathbf{N}$ is a field of unit vectors.

We can now deduce from (8)-(11) the following result

$$\int_{\Omega}(p - \gamma\mathrm{div}\mathbf{N})\mathbf{N}.\mathbf{m}ds = 0. \tag{13}$$

Since Ω and $\mathbf{m}$ are both arbitrary we have finally

$$p = \gamma\mathrm{div}\mathbf{N} \tag{14}$$

as the field equation governing the force balance at the outer surface of the liquid body.

Eq. (14) is rarely seen in engineering books or books on mathematical physics. These works mostly quote the famous capillarity law of Laplace

$$p = \gamma(\frac{1}{R_1} + \frac{1}{R_2}), \tag{15}$$

where R_1 and R_2 are the principal radii of curvature. Of course, the equality of (14) and (15) is well known in differential geometry.

The sessile-drop method to measure surface tension

When a small quantity of a liquid rests on a flat horizontal surface, it assumes a shape that may resemble that of Figs. 3a or 3b. The particular shape we will get depends upon the contact angle α. The value of the contact angle is determined by the three phases meeting one another at the contact line. One of these is almost always the surrounding air, the second being the particular liquid under consideration. The third phase is that of the substrate on which the drop rests. The liquid is said to be non-wetting in the case of Fig. 3a and wetting in that of Fig. 3b. In what follows we shall assume that we

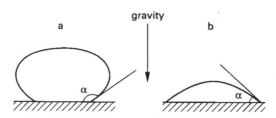

Figure 3: Sessile drop. Case a: Non-wetting, Case b: Wetting.

have to do with the case depicted by Fig. 3a, in which the drop has its largest width somewhere above the horizontal surface, not on it. We can always achieve this by choosing the right substrate material. As an example think of the surface of a car which has not been polished for quite a while. Rain falling on it will have a tendency to spread out. Once the car has had its polish, drops such as those of Fig. 3a will form and these may easily roll off.

It is possible to determine the surface tension of a liquid by measuring the shape of a

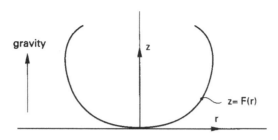

Figure 4: Coordinate system for sessile drop.

sessile drop such as that of Fig. 3a. Furthermore, it is possible to do so without having to know the value of the contact angle beforehand. In order to bring this about we turn the drop upside down and introduce a coordinate system as in Fig. 4. The coordinate z is antiparallel to the force of gravity. In the plane $z = 0$, which is tangent to the drop's apex, we have the cartesian coordinates x and y. However, the drop is axially symmetric, so that its shape can be described by a function such as

$$z = F(r), \tag{16}$$

where $r = \sqrt{x^2 + y^2}$.

The outward unit normals to the surface given by (16) are given by (a prime stands for differentiation with respect to r)

$$\mathbf{N} = \frac{1}{\sqrt{1 + (F')^2}} \frac{F'}{|F'|} \left(\frac{x}{\sqrt{x^2 + y^2}} F', \frac{y}{\sqrt{x^2 + y^2}} F', -1 \right), \tag{17}$$

where the factor $F' \mid F' \mid^{-1}$ ensures that we always have to do with the outward normal. It would seem that a difficulty arises in those points where F' changes sign. This happens where the drop reaches its maximum width. In that point F' changes from positive infinity to negative infinity. However, this difficulty is only a consequence of the way in which we describe the shape of the drop using Eq. (16). Had we used $r = f(z)$ this difficulty would not have arisen. However, since the literature mostly proceeds from an equation such as (16), we shall follow suit.

The pressure inside the drop is a simple hydrostatic one, i.e. p varies linearly with z as follows

$$p = p_o + \rho g z, \tag{18}$$

where ρ is the density (kg.m^{-3}), g the acceleration due to gravity (m.s^{-2}) and p_o the still unknown pressure at the apex inside the drop (kg.m^{-1}.s^{-2}). Using (14), (17) and (18) we can now derive the following differential equation for the shape of the drop

$$\frac{F'}{|F'|} \left\{ F'' + \frac{1}{r} F'(1 + (F')^2) \right\} = \frac{1}{\gamma} (p_o + \rho g F)(1 + (F')^2)^{\frac{3}{2}}. \tag{19}$$

This is a second-order nonlinear differential equation. The required drop shape can be obtained if we prescribe three conditions, namely two because the equation is of the second order and an additional one because of the unknown parameter p_o. Of course, without conclusive mathematical proof we cannot be sure that these three conditions suffice for determining the solution uniquely. However, we shall not go into this here.

Two boundary conditions are easily written down. At the apex we have

$$F = 0 \quad \text{and} \quad F' = 0 \quad \text{at} \quad r = 0. \tag{20}$$

The third condition is obtained from the measured shape of the drop. Suppose we measure a maximum radius of $r = a$, then our third condition will read

$$F' \to \infty \quad \text{when} \quad r \uparrow a. \tag{21}$$

Counting the number of parameters which determine this problem we come to four : γ, ρ, g and a. What we intend to do is compare an actual measured drop with all possible solutions of the system consisting of Eqs. (19)-(21) and then select the best fit. If all goes well, this yields an estimate of γ. The task confronting the applied mathematician is now to find a way of presenting all possible solutions of (19)-(21) in the most economical way. And again: dimensional analysis shows us the way.

Let us make this problem dimensionless by introducing the scaled variables ξ and η as follows

$$\xi = r/a, \quad \eta = z/a. \tag{22}$$

The problem is then formulated thus

$$\frac{\eta'}{|\eta'|} \left\{ \eta'' + \frac{1}{\xi} \eta' \left(1 + (\eta')^2 \right) \right\} = (\beta + Bo\eta) \left(1 + (\eta')^2 \right)^{\frac{3}{2}}, \tag{23}$$

where

$$Bo = \frac{\rho g a^2}{\gamma} \tag{24}$$

is the so-called Bond number. Furthermore, β is an unknown parameter which is related to the unknown pressure p_o. The boundary conditions read

$$\eta = 0, \eta' = 0 \quad \text{at} \quad \xi = 0 \tag{25}$$

and

$$\eta' \to \infty \quad \text{when} \quad \xi \uparrow 1. \tag{26}$$

Now, looking at the system consisting of Eqs. (23), (25) and (26) we come to the conclusion that all possible shapes can be classified as a one-parameter family of solutions. Indeed, for Bo and β given, the two conditions of Eq. (25) suffice to determine a unique solution. However, there is still Eq. (26) to contend with. This shows that the two parameters β and Bo are functions of one another. Once we have selected a value of Bo, the value of β is fixed. In other words: the Bond number uniquely determines the shape of the drop.

A numerical integration of the system (23), (25), (26) is not difficult to carry out. A minor technical difficulty arises on account of the change of sign of η' at $\xi = 1$. To circumvent this we have found it useful to rewrite (23) as a set of two equations for $\xi(s)$ and $\eta(s)$, where s is the arc length. We shall not give the details here. The end result is a series of graphs with Bo as the single parameter. Fig. 5 gives an impression of what these curves look like.

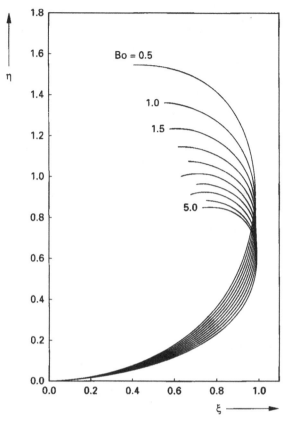

Figure 5: Normalised Bond curves.

Discussion

In principle, the measurement technique is very simple. A photograph is made of the sessile drop and, using this, its contour is measured as accurately as possible. The measured curve is blown up, or scaled down if need be, to such an extent that its maximum diameter is the same as that of the theoretical curves of Fig. 5. The measured curve is then compared with the theoretical ones, which leads to an optimal choice of the Bond number. Eq. (24) then tells us that γ follows from

$$\gamma = \frac{\rho g a^2}{Bo}. \tag{27}$$

Although this process may seem extremely simple, it is impossible to achieve a 100%-accurate curve fit yielding an equally accurate value of the Bond number. One source of error is clear: when measuring the contour of the drop, whether directly by means of an optical method or indirectly by means of a photograph, we incur errors of measurement. But apart from these obvious errors, there are other ones which are just as important.

The positioning of the substrate may deviate slightly from the horizontal, causing the drop to be asymmetrical. The surface tension may vary along the surface of the drop, owing to temperature gradients, compositional gradients or evaporation effects. This will give rise to Marangoni convection inside the drop, resulting in a shape different from the one following from the simple gravity-vs-surface-tension theory. A sensitivity analysis may reveal what influence each of these effects has on the shape of the drop and therefore on the value of the Bond number.

In conclusion, this small problem is another example showing how effective a process of non-dimensionalization can be. In the present case a single parameter, viz. *Bo,* would suffice for a complete description of the family of solutions. On the basis of this simple but fundamental theory we can build more complex theories which take into account the detrimental effects spelled out in the previous paragraph. Each single effect is represented by a dimensionless parameter. It is in the nature of these effects that these added-on parameters are small, allowing us to use perturbation methods to analyse their influence. However, this will be outside the scope of this presentation.

Suggestions for further reading

On the sessile drop method:

- C.A. Smolders and E.M. Duyvis, Receuil Trav. Chim. **80**, 635-649 (1961).

- D.N. Staicopolus, J. Coll. Interf. Sci. **17**, 439-447 (1962).

- J.N. Butler and B.H. Bloom, Surf. Sci. **4**, 1-17 (1966).

- Y. Rotenberg, L. Buruvka and A.W. Neumann, J. Coll. Interf. Sci. **93**, 169-183 (1983).

- S.H. Anastasiadis, J.-K. Chen et al., J. Coll. Interf. SCi. **119**, 55-66 (1987).

- H.K. Kuiken, Colloids and Surfaces **59**, 129-148 (1991).

Historical works:

- F. Bashforth and J.C. Adams, *An attempts to test the theories of capillary attraction* Cambridge University Press, Cambridge 1883.

- Kelvin (Lord), *Popular lectures and addresses,* I pp 1-72. London: The Royal Institution, 1886.

Mathematical works on capillarity:

- P. Concus and R. Finn, Phil. Trans. R. Soc. London **292** 307-340 (1979).

- R. Finn, *Equilibrium Capillary Surfaces* Springer, Berlin 1985.

- S.W. Rienstra, J. Eng. Math. **24** 193-202 (1990).

Mass transport from open cavities: an application to etching

Introduction: An etching experiment

Sometime during 1976 a colleague of mine approached me with a mass-transport problem. The man in question was someone whose task it was to produce a very fine metal gauze that was to be used in an optical experiment to be carried out in space. The holes in the gauze had to be more or less square and their depth had to be of the same order of magnitude as their widths. The idea was to produce this gauze by means of a so-called etching technique. To be able to get an insight into what this etching technique could do for him, he carried out a model experiment in which lines of various widths had to be etched in a metal sheet (Fig. 1). To make sure that the sheet was etched only according to this line pattern, the sheet was covered with a masking layer which prevented the etching of the parts of the sheet between the lines. The etchant, which is a corrosive liquid, could attack the exposed parts of the sheet and produce a two-dimensional hole pattern. In order to make this process as fast as possible, the etchant was splashed against the sheet with great force, the idea being that waste materials would be carried off as fast as possible, so that fresh etchant would always be present close to the surface.

This development scientist had found that, initially, etching proceeded at the same speed in every hole. However, at a certain point in time, the etching speed dropped markedly in the narrowest of the line-shaped holes. Somewhat later the same would happen with the smallest hole but one, and so on. By measuring the depths in the various holes at which the change in the etching speed would occur, our man found a constant depth-to-width ratio. This was the problem laid before me.

In the next few pages I shall describe how I explained the phenomenon.

Etching

Etching is a chemical process in which a substance dissolved in a liquid reacts with a solid, once the liquid and the solid have been brought into contact with one another. For instance a well-known substance for etching a metal such as iron is ferric chloride, $FeCl_3$, which can be dissolved in water up to very high concentrations. At the surface the Fe^{3+} ions are reduced to Fe^{2+} ions and the Fe atoms of the solid are oxidized to Fe^{2+} ions.

$$2Fe^{3+} + Fe \rightarrow 3Fe^{2+} \tag{1}$$

Once an Fe atom is oxidized, it leaves the surface and becomes part of the etchant, as a waste product, together with the reduced Fe^{2+} ions.

In the simplest model of such an etching process one considers only the concentration of the component which causes the etching process, which is the Fe^{3+}, and disregards the influence of the waste product. It is possible that at high enough concentrations of Fe^{2+}

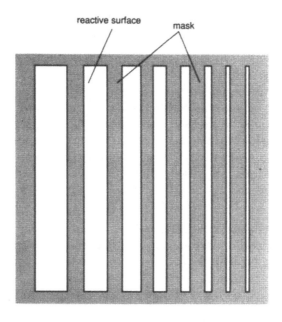

Figure 1: Etching a line pattern.

near the surface the etching process will be hampered considerably. This effect could be incorporated in a more refined model.

We introduce c as the concentration of the active etching component, i.e.

$$[c] = kmol.m^{-3} \tag{2}$$

where [ ] denotes the dimension. In this simple model one attempts to find the distribution of the concentration in the space of interest, which is the region occupied by the etchant.

What one wants to calculate, of course, is the etching speed. This quantity is related and indeed proportional to the amount of material that leaves the solid per unit of time. We have seen that this quantity, in turn, is proportional to the amount of dissolved active species which reaches the surface per unit time, which we denote by q. If n denotes the unit vector normal to the etching surface, then we can express this amount as

$$q = D\mathbf{n}. \text{ grad } c \text{ on } \Gamma , \tag{3}$$

where D is the diffusion coefficient of the dissolved active species with $[D] = m^2.s^{-1}$. Furthermore, Γ denotes the etching surface. On the other hand, it is well known from reaction kinetics that the amount of material which disappears in any chemical reaction is usually proportional to the concentration of that material itself. Therefore we have

$$q = kc \text{ on } \Gamma , \tag{4}$$

where k is the reaction constant with $[k] = m.s^{-1}$.
Combining (3) and (4), we obtain a boundary condition for our simple etching process

$$D\mathbf{n}. \text{ grad } c = kc \text{ on } \Gamma . \tag{5}$$

On the other hand, far from the etching surface the etchant is pure, so that the concentration of dissolved active species is that of the uncontaminated etchant

$$c = c_\infty \quad \text{on} \quad \Gamma_\infty \ , \tag{6}$$

where Γ_∞ denotes a surface infinitely far from the etching surface. Depending upon whether the chemical reaction proceeds quickly (k large) or slowly (k small), a limiting form of Eq. (5) can be used. If we let k tend to infinity, leaving the transport of etchant towards the surface unaltered, then the left-hand side of (5) will remain the same, but the right-hand side would seem to approach infinity, except, of course, when $c \downarrow 0$ at the surface. Such processes are called diffusion-controlled or, even better, transport-controlled. For such processes

$$c = 0 \quad \text{on} \quad \Gamma \ . \tag{7}$$

This simply means that an ion of the active species cannot exist on Γ during a finite time. Conversely when $k \downarrow 0$, leaving everything else unchanged, both terms in (5) will approach zero since c has a value bounded by (6). For such processes the concentration profile is flat, so that

$$c = c_\infty \quad \text{on} \quad \Gamma \ . \tag{8}$$

Such etching processes are called kinetically controlled. In this chapter we shall be concerned primarily with diffusion-controlled processes, so that (7) will be used throughout.

Returning now for a moment to the observed etching phenomenon which showed a marked decrease of the etch rate, we conclude that the gradient given by (3) integrated along the etched surface must have dropped rather suddenly. Let us suppose that the thickness of the layer across which the concentration drops from $c = c_\infty$ to $c = 0$ is given by δ. We conclude that the concentration gradient is of the order of

$$q = D\frac{c_\infty}{\delta} \ . \tag{9}$$

Since D and c_∞ are given parameters we are led to believe by this model that δ must have increased rather suddenly. It is not clear what physical phenomenon may have been the cause of this. Another explanation might be that the concentration drop across the layer is not c_∞ but some lower value. How this could be brought about is not clear either at this point.

Flow fields

When a liquid flows along a wall in which there is a depression, the liquid will penetrate the depression and shear along its bottom surface (Fig. 2a). When the depression becomes deeper, being more like a cavity, the exterior liquid will no longer penetrate all the way into the cavity, and the flow inside the hole will be a trapped eddy (Fig. 2b). This is true both for so-called creeping flows, in which the Reynolds number is very small, and for large-Reynolds-number flows in which inertia is the dominant factor. Multicellular flows may exist in deep cavities (Fig. 2c). Numerical studies have shown that the changeover from a no-cell situation to a one-cell situation occurs rather suddenly without a large change of shape being needed to effectuate this change-over.

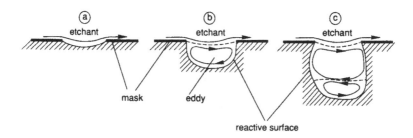

Figure 2: Possible flow fields within holes.

The Schmidt number

In the etching experiment described in the introduction the motion of the etchant played an important role. Indeed, to speed up the experiment, the etchant was splashed with great force against the surface to be etched. Then, apart from diffusional transport, epitomized by the diffusion coefficient D, there is convection to be dealt with. We shall assume that the flow field is given and that it is not directly influenced by the concentration c. Of course, there is an indirect influence since the etching process will make the cavity deeper and deeper, thus changing the geometry continually. However, with the geometry given at any point in time, assuming that the flow field changes in a quasi-stationary manner, we can calculate this flow field during the entire etching process.

It is important to know how the velocity gradients relate to the concentration gradients. Just as the latter are determined to a large extent by the diffusion coefficient D, so are those of the velocity determined by the kinematic viscosity ν. For ν we have $[\nu] = m^2 s^{-1}$ which is the same dimension as that of D. A simple example will show how these two relate to one another.

Let us assume that we have a semi-infinite expanse of viscous liquid occupying the region $y > 0$. This infinite expanse is bounded by a plane wall occupying the entire plane $y = 0$. Both the liquid and the plane move uniformly in a given direction at a velocity u_0. At a given time $t = 0$ the motion of the plane is suddenly brought to a halt. Since the liquid is viscous, the no-slip condition prevails at the wall, so that along with the wall the liquid at the wall is also brought to a stand-still. It is well-known that the subsequent motion of the liquid is given by

$$u = u_0 \text{erf}\left\{ y/(2\nu t)^{\frac{1}{2}} \right\} \ . \tag{10}$$

Let us suppose next that the liquid is an etchant in which the concentration of the active etching component is given by c_∞. At $t = 0$ the condition of Eq. (7) suddenly comes into effect. The subsequent concentration profiles will be given by

$$c = c_\infty \text{erf}\left\{ y/(2Dt)^{\frac{1}{2}} \right\} \ . \tag{11}$$

The error function approaches its asymptotic value, which is equal to unity, exponentially fast. Therefore the thicknesses of the velocity layer and the concentration layer are given

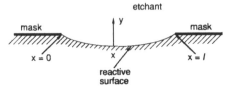

Figure 3: Coordinate system for shallow cavity.

by

$$\delta_u \sim (2\nu t)^{\frac{1}{2}} \quad \text{and} \quad \delta_c \sim (2Dt)^{\frac{1}{2}} \tag{12}$$

respectively, from which we have

$$\frac{\delta_u}{\delta_c} = \left(\frac{\nu}{D}\right)^{\frac{1}{2}} = (Sc)^{\frac{1}{2}} \tag{13}$$

where

$$Sc = \frac{\nu}{D} \tag{14}$$

is the Schmidt number, which is dimensionless. For liquids Sc is usually very much larger than unity. Indeed, for a liquid such as water we have $\nu \sim 10^{-6} m^2 s^{-1}$ and a typical value of D is $10^{-10} m^2 s^{-1}$. Therefore, the concentration layer is very much thinner than the velocity layer. Indeed, within the extremely thin concentration layer the velocity layer could easily be approximated by

$$u \sim u_0 \frac{2}{\pi^{\frac{1}{2}}} \frac{y}{(2\nu t)^{\frac{1}{2}}} \quad , \tag{15}$$

where we have written down the first term of the Taylor series expansion of the error function. This observation is only of minor importance in this simple model problem, the reason being that the concentration distribution is not influenced by the velocity field. However, we can make use of it in the convection-diffusion problem that we shall discuss later, since the motion of the etchant is certainly of influence on the concentration there.

Mass transport from a cavity without an eddy

Let us assume that the etched cavity is still shallow enough for the flow field to be that of Fig. 2a. We shall introduce a coordinate system fitted to the cavity wall as in Fig. 3. Let x measure distance along the cavity wall and y distance from the wall along the normal. Then, excluding higher-order effects which play a role only when the wall is sharply curved, we have to consider the equation

$$u\frac{\partial c}{\partial x} + v\frac{\partial c}{\partial y} = D\frac{\partial^2 c}{\partial y^2} \quad , \tag{16}$$

where u and v are the velocity components in the x and y directions respectively. Eq. (16) is a so-called boundary-layer equation. It is valid for the situation considered by us as long as the concentration layer is thin, i.e. much thinner than the length of the cavity. This condition prevails as long as the Peclet number

$$Pe = \frac{u_c l}{D} \tag{17}$$

is much larger than unity. In (17) u_c is a typical velocity *within* the layer and l is the length of the cavity. Since, as we have seen, D is very small, Pe is usually far greater than unity.

We shall now make use of the result of our discussion on the Schmidt number. Within the boundary layer the velocity field can be approximated by the first term of a Taylor-series expansion. This means

$$u \sim \alpha(x)y , \quad v = -\frac{1}{2}\alpha'(x)y^2 , \tag{18}$$

where a prime stands for differentiation with respect to argument. The meaning of the Taylor-series term for u should be clear: u must be zero on the wall ($y = 0$); the normal gradient of u varies along the wall ($\alpha(x)$). The approximation for v then follows from the equation of continuity, which is $\partial u/\partial x + \partial v/\partial y = 0$, and the condition $v = 0$ at $y = 0$.

The boundary conditions are

- at $x = 0$ fresh uncontaminated and unspent etchant arrives at the cavity:

$$c = c_\infty \quad \text{at} \quad x = 0 (y > 0) , \tag{19}$$

- at the cavity wall the diffusion-controlled etching limit prevails:

$$c = 0 \quad \text{at} \quad y = 0 \quad (0 < x < l) , \tag{20}$$

- at the outer edge of the boundary layer the etchant is unspent:

$$c \to c_\infty \quad \text{when} \quad y \to \infty . \tag{21}$$

The condition $y \to \infty$ should not be taken literally, since we consider finite systems. The solutions we find vary in an exponential sense in such a way that the outer conditions are satisfied with great accuracy at only a short distance from the wall.

To solve this problem we apply a very effective transformation that was discovered by Lighthill. We introduce transformed dimensionless variables θ, δ and η as follows:

$$\theta(\xi, \eta) = c_\infty^{-1} c(x, y) , \tag{22}$$

$$\xi = \xi_0^{-1} \int_0^x \alpha^{\frac{1}{2}}(\tilde{x}) d\tilde{x} , \tag{23}$$

$$\eta = \alpha^{\frac{1}{2}}(x)(D\xi_0)^{-\frac{1}{3}} y , \tag{24}$$

with

$$\xi_0 = \int_0^l \alpha^{\frac{1}{2}}(\tilde{x}) d\tilde{x} . \tag{25}$$

The resulting problem now reads

$$\eta \frac{\partial \theta}{\partial \xi} = \frac{\partial^2 \theta}{\partial \eta^2} \ , \tag{26}$$

$$\xi = 0(\eta > 0) : \theta = 1 \ , \tag{27}$$

$$\eta = 0(0 < \xi < 1) : \theta = 0 \ , \tag{28}$$

$$\eta \to \infty : \theta \to 1 \ . \tag{29}$$

The amazing result of Lighthill's transformation is that the velocity field no longer appears explicitly in the problem definition of Eqs. (26)-(29). It is hidden in the transformation of Eqs. (22)-(25). Therefore, we can solve this problem without having to know the velocity field explicitly. Only when we need explicit numerical values, do we have to give the velocity field also.

The problem defined by (26)-(29) is classical. It is easily solved by means of the Laplace transform. Alternatively, we could observe that the transformation

$$\theta(\xi, \eta) \to \overline{\theta}(\overline{\xi}, \overline{\eta}) \quad \eta \to \overline{\eta} K^{\frac{1}{3}} \quad \xi \to \overline{\xi} K \ , \tag{30}$$

where K is an arbitrary constant, leaves the problem definition unchanged. Therefore, the solution can be a function of

$$\sigma = \eta \xi^{-\frac{1}{3}} \tag{31}$$

only. Writing

$$\theta = f(\sigma) \ , \tag{32}$$

we find

$$f'' + \frac{1}{3}\sigma^2 f' = 0 \ ; \quad f(0) = 0 \ , \quad f(\infty) = 1 \ , \tag{33}$$

which has the solution

$$\theta(\xi, \eta) = f(\sigma) = \frac{\gamma(\frac{1}{3}, \frac{\sigma^3}{9})}{\Gamma(\frac{1}{3})} \ , \tag{34}$$

where

$$\gamma(a, b) = \int_0^b q^{a-1} e^{-q} dq \tag{35}$$

is the incomplete Gamma function and Γ denotes the (complete) Gamma function with $\Gamma(b) = \gamma(\infty, b)$.

An important result is the mass-transport rate which is defined by

$$Q = \int_0^l q(x) dx = D \int_0^l \frac{\partial c}{\partial y}|_{y=0} dx = \frac{3^{\frac{1}{3}}}{\Gamma(\frac{1}{3})} c_\infty \left(\frac{D^2}{\int_0^l \alpha^{\frac{1}{2}}(\sigma) d\sigma} \right)^{\frac{1}{3}} \int_0^l \frac{\alpha^{\frac{1}{2}}(x) dx}{(\int_0^x \alpha^{\frac{1}{2}}(\sigma) d\sigma)^{\frac{1}{3}}} \tag{36}$$

where we have used (3), (22)-(25), (31), (34) and (35). This quantity clearly depends upon the velocity field, namely through the function α. When the velocity gradients at the surface increase, for instance by means of an enhanced stirring level in the etchant, Q increases also. When α increases by a multiplicative factor p all along the etching surface, Q increases by the factor $p^{\frac{1}{3}}$. It is also seen that Q is directly proportional to c_∞, which is the same as the result we obtained from a simpler model (9).

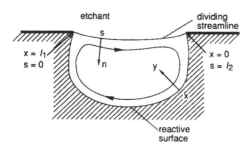

Figure 4: Coordinates for single-eddy situation.

Mass transfer from a cavity with a trapped eddy

Let us now consider a situation as sketched in Fig. 2b. The cavity is deeper now so that the outside etchant does not penetrate it all the way to the bottom. There is a streamline near the mouth of the cavity which separates the flow outside from that inside. Now suppose that the flow inside the cavity is still strong enough for the Peclet number (17) to be much larger than unity. As an example let us consider a small cavity of the size of one hundred microns (10^{-4} metres). Then with $D \sim 10^{-10}m^2/s$ we conclude that the velocities inside the cavity should be considerably (at least 100 times, say) larger than $10^{-6}m/s$. This would mean velocities of the order of 100 microns per second which is not overly large. Therefore within a cavity of 100 microns large values of Pe are easily attained.

When $Pe \gg 1$, mass transfer will occur in boundary layers along the rim of the eddy as shown in Fig. 4. One part of this boundary layer, namely the part along the cavity wall of length l, is of the kind considered in the previous section. The remaining part is along the separating streamline. The wall boundary layer, which was in direct contact with the outer etchant in the previous section, is now screened off from the outer etchant by the free- streamline boundary layer. The active etching component must first pene-trate, i.e. diffuse through, the free-streamline boundary layer, before it can enter the cavity region. In order for this etchant to be able to diffuse through the free-streamline boundary layer, there must be a concentration difference across it. Therefore we infer that the concentration of active etching component within the cavity, but outside the two boundary layers, i.e. in the region surrounded by the boundary layers, must be a constant below c_∞, let us say c_i, with $0 < c_i < c_\infty$.

We are now in a position to explain the drop in the etch rate. We have seen in the previous section that the etch rate is directly proportional to the concentration difference across the wall boundary layer. When there is a trapped eddy within the cavity, this difference will be c_i instead of c_∞. Also, the velocities are expected to become smaller as the cavity becomes deeper. We have seen that this has a small, but still lowering, effect upon the etch rate.

Mathematical model for the etching of a relatively deep cavity

We shall now devise a mathematical model for the single-eddy situation. What we must do is to find solutions for two different boundary-layer problems and patch these at either end. The wall-boundary-layer problem is almost the same as that expressed by Eqs. (16), (18)-(21) except that c_∞ has to be replaced by c_i, where c_i is a still unknown constant, and where condition (19) has to be replaced by

$$c = f(y) \text{ at } x = 0 (y > 0) \ , \tag{37}$$

where $f(\infty) = c_i$. The function $f(y)$ is still unknown and will be determined as part of the solution process. To solve this problem we shall apply a transformation similar to the one given by Eqs. (22)-(25):

$$\theta(\xi, \eta) = c_i^{-1} c(x, y) \tag{38}$$

where ξ and η are defined by (23)-(25). Furthermore, we shall assume that the length of the cavity wall is equal to l_1. Condition (37) is replaced by

$$\theta = F(\eta) \text{ at } \xi = 0 (\eta > 0) \tag{39}$$

where $F(\infty) = 1$.

The solution to this problem, which is derived in the Appendix, is

$$\theta(\xi, \eta) = \frac{1}{\Gamma(\frac{1}{3})} \gamma\left(\frac{1}{3}, \frac{\eta^3}{3\xi}\right) +$$
$$+ \frac{1}{3} \frac{\eta^{\frac{1}{2}}}{\xi} \exp(-\eta^3/9\xi) \int_0^\infty \tau^{\frac{1}{2}} (F(\tau) - 1) \exp(-\tau^3/9\xi) I_{\frac{1}{3}}(\frac{2}{9}\tau^{\frac{3}{2}}\eta^{\frac{3}{2}}/\xi) d\tau \ , \tag{40}$$

where $I_{\frac{1}{3}}$ is a modified Bessel function.

The solution to the free-streamline boundary-layer problem can be derived in a similar fashion. The flow field within the free-streamline region can be approximated by

$$u(s, n) \sim \beta(s) \text{ and } v(s, n) \sim -\beta'(s)n \ , \tag{41}$$

where s and n are the coordinates of a locally orthogonal coordinate system with $n = 0$ on the free streamline, and the streamline itself being defined by $0 < s < l_2$. The governing differential equation is again Eq. (26). The boundary conditions are

$$c = g(n) \text{ at } s = 0 \ , \tag{42}$$

with $g(-\infty) = c_\infty$ and $g(\infty) = c_i$, and

$$c \to c_\infty \text{ when } n \to -\infty \ , \tag{43}$$

$$c \to c_i \text{ when } n \to \infty \ . \tag{44}$$

This problem can be simplified considerably by the application of a transformation similar to the one proposed by Lighthill

$$\vartheta(\mu, \nu) = c_\infty^{-1} c(s, n) \ , \tag{45}$$

$$\mu = \mu_0^{-1} \int_0^s \beta(\breve{s}) d\breve{s} \quad , \tag{46}$$

$$\nu = \beta(s)(D\mu_0)^{-\frac{1}{2}} n \quad , \tag{47}$$

$$\mu_0 = \int_0^{l_2} \beta(\breve{s}) d\breve{s} \quad . \tag{48}$$

The problem is now reduced to

$$\frac{\partial \vartheta}{\partial \mu} = \frac{\partial^2 \vartheta}{\partial \nu^2} \quad (0 < \mu < 1 \, , \quad -\infty < \nu < \infty) \tag{49}$$

$$\mu = 0 (\nu > 0) : \vartheta = G(\nu) \quad , \tag{50}$$

$$\mu = 0 (\nu < 0) : \vartheta = 1 \quad , \tag{51}$$

$$\nu \to -\infty : \vartheta \to 1 \quad , \tag{52}$$

$$\nu \to \infty : \vartheta \to \vartheta_i = c_i/c_\infty \quad . \tag{53}$$

Condition (51) expresses the fact that fresh etchant is supplied from the outside.
The solution to this problem can easily be derived by means of the Laplace transform and reads

$$\vartheta(\mu, \nu) = \frac{1}{2} \mathrm{erfc} \left(\frac{\nu}{2\mu^{\frac{1}{2}}} \right) + \frac{1}{2(\pi\mu)^{\frac{1}{2}}} \int_0^\infty G(\tau) \exp \left(-\frac{(\nu - \tau)^2}{4\mu} \right) d\tau \quad . \tag{54}$$

It now remains for us to determine the unknown functions $F(\eta)$ and $G(\nu)$. Indeed, once these functions are known, we shall be able to evaluate the expressions (40) and (54) numerically. Obviously, as the two boundary layers bite each other's tails, the function $F(\eta)$ is related to the end value of Eq. (54) at $\mu = 1$ and, similarly, $G(\nu)$ is related to $\vartheta(1, \nu)$. Strictly speaking, boundary-layer analyses are invalid in regions of abrupt changes. Therefore an exact matching is out of the question and we have to settle for something less exact. What we shall assume is that the concentration, i.e. the values of θ and ϑ, are constant along streamlines in these regions of abrupt change. Streamlines are defined by the function

$$\Psi = \int_0^y u(x, \overline{y}) d\overline{y} \quad . \tag{55}$$

In the case of the wall boundary layer this gives

$$\Psi = \frac{1}{2} \alpha(x) y^2 \tag{56}$$

and for the free-streamline boundary layer

$$\Psi = \beta(s) n \quad . \tag{57}$$

Patching occurs at the coordinate pairs $(x, s) = (l_1, 0)$ and $(x, s) = (0, l_2)$. Our earlier assumption is now tantamount to the equality of the expressions of the stream function and the concentrations at these two points. Let us first look at the stream functions. Eqs. (56) and (57) at the two points lead to the expression

$$\nu = \omega \eta^2 \tag{58}$$

with

$$\omega = \frac{1}{2} \frac{(D\xi_0)^{\frac{2}{3}}}{(D\mu_0)^{\frac{1}{2}}} \; , \tag{59}$$

where we have used (24) and (47). Eq. (58) relates the two coordinates ν and η at the two boundary-layer transition points. Applying this to (40) and (54), we can now write

$$\theta(0,\eta) = \vartheta(1,\omega\eta^2) \text{ and } \theta(1,\eta) = \vartheta(0,\omega\eta^2) \; . \tag{60}$$

These two conditions lead to two coupled integral equations for the two unknown functions $F(\eta)$ and $G(\nu)$. It is now a matter of applying the right numerical analysis to find the solution. As we are concerned here solely with mathematical modelling, we shall not present the details here. The interested student may consult the list of suggested reading for further information on this topic.

Concluding remarks

In this chapter we have seen how one can explain a certain physical phenomenon by bringing together different physical effects and combining those into a simple mathematical model. The physical ingredients were
   a) diffusion-controlled etching;
   b) flow fields in open cavities;
   c) boundary layer theories for large Schmidt numbers.
Although the model eventually led to a non-trivial system of two coupled integral equations, the explanation of the phenomenon, namely the sudden drop in the etch rate, could already be given in a qualitative sense on the basis of a thorough understanding of a), b) and c).

Of course, it is one thing to explain a phenomenon, it is quite another to come up with a solution on how to prevent it from occurring. The drop in the etch rate is detrimental when holes of different sizes have to be etched simultaneously. The etch rate drops when a certain depth-to-width ratio is reached, so that the absolute depths at which this occurs depends upon the size of the cavity. Therefore, the times needed to etch different holes vary with the hole size. For certain applications this may be an undesirable effect.

Obviously, one way to prevent this kind of etching is to make sure that the trapped-eddy situation does not occur. One way of doing that is to etch in a gravitational or better centrifugal field. We refer to the references for further information.

It is perhaps worthwhile to say something here about how one should present a paper in which the mathematical modelling of a physical phenomenon is the central theme. Quite understandably, mathematicians have a tendency to highlight the mathematics in their presentations. However, in papers of the kind alluded to above, and in this chapter we saw one of them, the emphasis should be on the translation of physical ideas into mathematics. Therefore, these papers ought to contain sizeable verbal discussions. As soon as the main theme is in danger of being lost by the need to present detailed mathematical calculations or paraphernalia that are otherwise of a technical mathematical nature, it is better to present these in an appendix. This is what we have done here. This is quite contrary to ordinary mathematical practice where there is a tendency to emphasize the 'better' mathematics. In industrial/engineering mathematics the mathematics serves some purpose outside mathematics. It is not an end in itself.

Appendix

Let us consider the problem consisting of the differential equation

$$\eta\frac{\partial\theta}{\partial\xi} = \frac{\partial^2\theta}{\partial\eta^2} \quad (\xi > 0, 0 < \eta < \infty) \tag{A1}$$

and the boundary conditions

$$\xi = 0(\eta > 0): \theta = F(\eta), \tag{A2}$$

$$\eta = 0(\xi > 0): \theta = 0, \tag{A3}$$

$$\eta \to \infty: \theta \to 1. \tag{A4}$$

We can write θ as the sum of two functions

$$\theta = \theta_1 + \theta_2 , \tag{A5}$$

where θ_1 is given by Eq. (34). It satisfies (A1), (A3) and (A4). At $\xi = 0(\eta > 0)$ we have $\theta_1 = 0$. The function θ_2 satisfies homogeneous boundary conditions at $\eta = 0$ and for $\eta \to \infty$ and also condition (A2). To solve the problem for θ_2, let us replace $F(\eta)$ by $\delta(\eta - \eta_s)$ where δ represents the delta function. Applying the Laplace transform

$$\bar{\theta} = \int_0^\infty e^{-p\xi}\theta(\xi,\eta)d\xi , \tag{A6}$$

we obtain the transformed problem

$$\frac{d^2\bar{\theta}_2}{d\eta^2} - \eta\{p\bar{\theta}_2 - \delta(\eta - \eta_s)\} = 0 \tag{A7}$$

with

$$\bar{\theta}_2(0) = 0 , \quad \bar{\theta}_2(\infty) = 0 . \tag{A8}$$

The general solution of the homogeneous equation of (A7) can be expressed as follows

$$\bar{\theta}_2 = \alpha\eta^{\frac{1}{2}}K_{\frac{1}{3}}\left(\frac{2}{3}p^{\frac{1}{2}}\eta^{\frac{3}{2}}\right) + \beta\eta^{\frac{1}{2}}I_{\frac{1}{3}}\left(\frac{2}{3}p^{\frac{1}{2}}\eta^{\frac{3}{2}}\right) , \tag{A9}$$

where $K_{\frac{1}{3}}$ and $I_{\frac{1}{3}}$ are modified Bessel functions. Using the well-known method of variation of parameters (here α and β), we can obtain the solution to the full problem (A7)-(A8). This reads

$$\bar{\theta}_2 = \frac{2}{3}\eta_s\eta^{\frac{1}{2}}K_{\frac{1}{3}}\left(\frac{2}{3}p^{\frac{1}{2}}\eta_s^{\frac{3}{2}}\right)I_{\frac{1}{3}}\left(\frac{2}{3}p^{\frac{1}{2}}\eta^{\frac{3}{2}}\right) , \quad (\eta < \eta_s)$$

$$= \frac{2}{3}\eta_s\eta^{\frac{1}{2}}I_{\frac{1}{3}}\left(\frac{2}{3}p^{\frac{1}{2}}\eta_s^{\frac{3}{2}}\right)K_{\frac{1}{3}}\left(\frac{2}{3}p^{\frac{1}{2}}\eta^{\frac{3}{2}}\right) , (\eta > \eta_s) . \tag{A10}$$

Using formula 13.96 of F. Oberhettinger & L. Badii, Tables of Laplace Transforms, which reads

$$L^{-1}\left\{I_{\frac{1}{3}}(ap^{\frac{1}{2}})K_{\frac{1}{3}}(bp^{\frac{1}{2}})\right\} = \frac{1}{2\xi}\exp\left\{-\frac{1}{4}\frac{a^2+b^2}{\xi}\right\}I_{\frac{1}{3}}\left(\frac{ab}{2\xi}\right), \tag{A11}$$

we find

$$\theta_2 = \frac{1}{3\xi}\eta_s\eta^{\frac{1}{2}}I_{\frac{1}{3}}\left(\frac{2}{9}\frac{\eta_s^{\frac{3}{2}}\eta^{\frac{3}{2}}}{\xi}\right)\exp\left\{-\frac{1}{9\xi}(\eta_s^3+\eta^3)\right\}. \tag{A12}$$

This solution applies when $F(\eta)$ is replaced by $\delta(\eta - \eta_s)$. The solution which satisfies (A2) for a general function $F(\eta)$ is

$$\theta_2 = \frac{\eta^{\frac{1}{2}}}{3\xi}e^{-\frac{\eta^3}{9\xi}}\int_0^\infty \eta_s^{\frac{1}{2}}F(\eta_s)\,e^{-\frac{\eta_s^3}{9\xi}}I_{\frac{1}{3}}\left(\frac{2}{9}\frac{\eta_s^{\frac{3}{2}}\eta^{\frac{3}{2}}}{\xi}\right)\mathrm{d}\eta_s. \tag{A13}$$

Suggestions for further reading

A more detailed presentation of the problem can be found in:

- H.K. Kuiken, Heat or mass transfer from an open cavity, J. Eng. Math. **12**, 129-155 (1978).

On flows in open cavities and cellular flows:

- H.K. Moffatt, J. Fluid Mech. **18**, 1-18 (1964).

- V. O'Brien, Phys. Fluids **15**, 2089-2099 (1972).

- F. Pan & A. Acrivos, J. Fluid Mech. **28**, 643-655 (1967).

- M. Takematsu, J. Phys. Soc. Japan **21**, 1816-1821 (1966).

- S.A. Trogdon and D.D. Joseph, J. Non-Newt. Fluid Mech. **10**, 185-213 (1982).

On the theory of diffusion-controlled transport:

- V.G. Levich, *Physico-chemical Hydrodynamics* Prentice Hall, Englewood Cliffs, 1962.

- H.R. Thirsk & J.A. Harrison, *A guide to the study of electrokinetics.* Academic Press, London, 1972.

On centrifugal and gravitational etching:

- H.K. Kuiken & R.P. Tijburg, J. Electrochem. Soc. **130**, 1722-1729 (1983).

On Lighthill's transformation:

- M.J. Lighthill, Proc. R. Soc. London **A202**, 369-377 (1950).

- A. Acrivos, Phys. Fluids **3**, 657-658 (1960).

On the Schmidt and the Peclet number:

- W.J. Beek & K.M.K. Muttzall, *Transport phenomena*. Wiley, London, 1975.

INVERSE PROBLEMS IN MATHEMATICS
FOR INDUSTRY

Bruno Forte
Department of Applied Mathematics
University of Waterloo
Waterloo, Ontario, Canada   N2L 3G1

1. INTRODUCTION

Due mainly to the lack of data most of the mathematical problems that arise in modelling an industrial process are "inverse problems". In these lectures we shall present a few real life problems that lead to different kind of inverse problems. The discussion of some related mathematical problems is also on the agenda. The aim is to achieve a better understanding of what an inverse problem is. We also hope that at the end an answer can be found to the major questions: how to formulate an inverse problem when modelling a real life situation. Let us first try to find a definition for "inverse problem."

Several attempts have been made to define "inverse problems", in various fields: classical mechanics, elasticity, partial differential equations, dynamical systems, etc.

G.K. Suslov (1890) [21] says "By inverse problem of **mechanics** we mean the determination of forces from the given properties of motion". According to A.S. Galiullin (1984) [8] "the possibility of modelling a number of applied problems ...has lead to a considerable broadening of the concept of inverse problems during the past few years. Problems have been formulated not only to determine the generalized forces, but also the parameters of a mechanical system, resulting in the motion of a mechanical system with the given properties". G.M.L. Gladwell (1989) [12] suggests the following definition "Given some well-known or traditional problem, an inverse problem (of this problem) can be viewed (loosely) as a problem in which some or all of the given quantities (**input data**) have been interchanged with some or all the required quantities (**output data**)".

A more general (and abstract) definition of inverse problem can be derived by generalizing a remark made by M.M. Lavrentiev (1967) [17], in defining "properly posed" problems.

Let Φ, F be some complete metric spaces, and let A be a function with domain Φ and range F. Consider the equation

$$A\phi = f \tag{1.1}$$

according to Lavrentiev "most of the problems of mathematical physics can be reduced to the investigation of the solution ϕ of equation (1.1) with a **given** function A and right-hand side f".

To generate a mathematical model of a real life situation one has indeed to define the spaces Φ and F, then produce the functions A and f. To solve the mathematical (direct) problem one has to find ϕ, that is one has to find a function Bf (the inverse of A) such that $\phi = Bf$. Any problem where the unknown is A and/or f, or part of them, while some features of ϕ are given, is an inverse problem.

We can say in a few words that in order to create a mathematical model one has to state and solve an inverse problem. This is why inverse problems extend over a huge area for research.

In the next section we shall show how modelling some real life situation requires the solution of (unsolved) inverse problems. Then we shall be dealing with three typical inverse problems in classical mechanics and their formal solution. We will close these lectures with an application of generalized dynamical systems to solve a "visual" inverse problem: the representation of a given image by a suitable algorithm.

2. EXAMPLES OF INVERSE PROBLEMS IN INDUSTRY

We have selected as examples three problems with three different degrees of incompleteness of the data, which are necessary to define the functions A and f.

Ex. 1 - Semi-automation in operating a locomotive engine: Full automation in operating train, airplanes, cable-cars, etc., has always been one of the objectives of the management of the respective companies. Full automation has been, indeed, successfully implemented by airlines, subway transit systems, railroad companies (limited to passenger trains). Serious difficulties arise when dealing with freight trains. This is mainly due to the heterogeneity in the composition of the train which normally consists of a series of different freight cars: boxcars, bedcars, tankcars, etc., unevenly loaded; unlike subway trains, aircraft, passenger trains where uniformity is the rule. For this reason (and probably some other reasons) the railroad companies have opted for a partial automation. In the semi-automated operation the engineer is assisted by a computerized system that will take over operation only in case of emergency, more specifically only when the engineer has failed to take action in due time, as needed, to stop the train. The computerized system takes over as soon as the actual distance of the locomotive from the point where it is supposed to stop exceeds the so-called "limit of authority". The limit of authority for each value v of the train speed is the distance required by the locomotive that pulls it to come to a full stop from the initial value v by full application of the brakes.

Thus, a model is needed to provide for each train and each value v of its speed the limit of authority. Theoretically, Newton's law defines the structure of equation (1.1) which is that of an initial value problem for a dynamical system. Unfortunately the essential parameters (masses, external and internal forces, frictional forces, etc.) that enter into the explicit representation of equation (1.1) are largely unknown. So we are unable to produce a model and solve the problem as a (**direct**) problem of classical mechanics. In fact, when the train leaves the terminal the engineer knows only the total number of cars, the total power of his engine and possibly the gross total weight of the cars. In order to find a feasible model, one can examine what the engineer does when he tries to become acquainted with the braking capabilities of its train. He, in fact, "tries the brakes" by a partial application and "feels" the reaction of the train to this light braking. From this he extrapolates (following his experience) the behaviour of the train consequent to a full brake application. This kind of (input) data, i.e. decelerations curves for various partial brake applications (see Fig. (2.1)), are not only available but quite reliable.

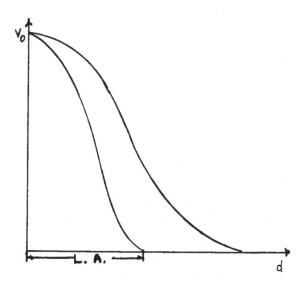

Figure 2.1

Modelling by this natural approach requires the solution of the following inverse problem: given a certain number of deceleration curves, i.e. a certain number of solutions ϕ of equation (1.1) find the functions A and f and then solve equation (1.1) for each initial value of the train velocity to determine the corresponding limit of authority (a direct problem). To the best of my knowledge this inverse problem is still unsolved.

It would be even better if without going through equation (1.1), whose knowledge is redundant, one could find directly the required deceleration curves (output data for full brake applications) from a certain number of deceleration curves (partial brake applications). This, indeed, the engineer does, resorting to his own past "experience".

Ex. 2 - Search techniques for unmanned air vehicles: In the direct problem a solution of equation (1.1) is given so that the motion of the aircraft is known. The relative motion of the camera installed on the plane is also known as is known the rate of picture taking. While the plane flies over the area S, to be searched, a sequence of small areas s_i is generated (Fig. (2.2)), s_i (footprint) being the area covered by the i-th picture. The search operation ends when the footprints have covered the whole area S.

The practical problem is an inverse problem with respect to the one that we have just described.

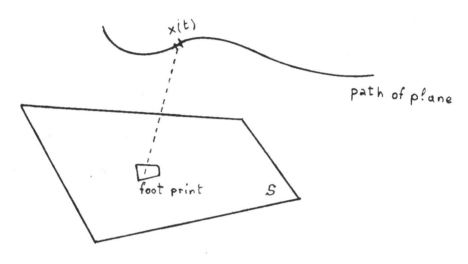

Figure 2.2

In practice one wants to know, given the area S to be searched, how to program the controls of aircraft and camera so that the corresponding sequence of footprints covers S. For this inverse problem the structure of A and f can be found (Newton's law).

Some of the "forces" (the controls) are not given, they are the unknowns in the inverse problem. As stated this inverse problem does not have a unique solution. So one can be looking for an optimal solution according to some criterion of optimality like minimum time, minimum fuel consumption etc. What makes this problem rather complex is the presence of several constraints (on the tilt rate, on the yaw, pitch, roll and pan rates of the aircraft, for example). Again, to the best of my knowledge none of the above mentioned problems has been fully solved.

This example and example 1 fall into the general area of classical mechanics. In paragraph 3 we will show three types of inverse problems in classical mechanics. They cover more theoretical than practical situations. The interesting aspect is that for them general methods to find solutions are available. The need for more research to reduce the gap between theory and practice will be evident.

Ex. 3 - The batch coil annealing process: The batch annealing is a stage in processing sheets of steel after the so-called cold rolling. Steel sheets come out of the cold rolling in the form of rolls, each roll consists of an average of 130 spirals. The purpose of the process is to restore the steel's ductility and flexibility, rebuilding its internal grain structure. This is attained by heating the steel up to a certain temperature (process that requires about 30 hours), then allow it to cool slowly (stage that takes usually almost twice as much time). A mathematical model of the batch annealing has the following objectives:

1- To reduce the anneal time while preserving and eventually improving the quality of the steel. This can be achieved by experimenting different technical solutions on the model.

2- To provide a better understanding of the heat transfer throughout the coils (the steel rolls).

3- To try different schedules for switching on and off the furnace, in order to find the optimal one and save energy.

4- To create a data file for future full automation of the annealing process.

A typical installation for the batch annealing can be described as follows. Four stacks of coils are placed on a plateau. Each stack consists of 2-5 coils and sits on a bottom convector that allows protective gas to flow through. A cover is placed on each stack to keep the stack in a controlled non-oxidizing gas atmosphere. The gas is kept circulating by a fan of radial type in the plateau underneath the bottom convector. Separating plates (convectors) are placed between the coils to let the gas circulate all around the coils (see Fig. (2.3)).

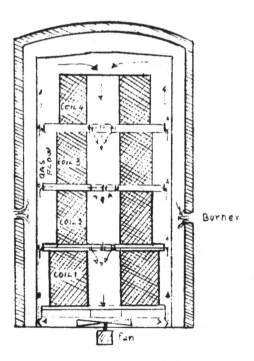

Figure 2.3

Finally a furnace is placed on the plateau to enclose the four stacks with their covers (see Fig. (2.4)). Eight burners are installed in the furnace (four on each side of the stacks).

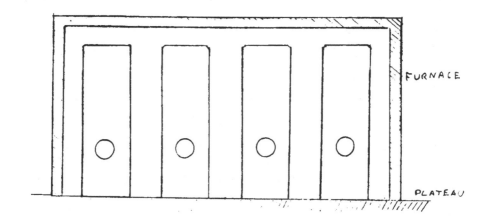

Figure 2.4

The gas has a triple role:

i) it prevents oxidization,

ii) it helps reaching a more uniform temperature distribution over the coils,

iii) it provides a source of heat for the inner surface of the coils, allowing for more heat transfer across the coils (actually the heat transfer along the axial direction of the coil can be five times that across the coil).

In a simplified mathematical model, that assumes as given the temperature distribution on the cover, at each point in time, the modes of heat transfer are,

1. *Gas atmosphere*   Radiation from coils and cover, convection from all surfaces contacted.

2. *Convectors*   Convection from gas, conduction to coil ends.

3. *Coil outside surface*   Radiation from cover and to gas, convection from gas.

4. *Coil inside surface*   Convection from gas, radiation to gas.

5. *Coil ends*   Convection from gas, conduction from convector, at top of top coil radiation from cover and to gas.

In addition the heat flow by the motion of the gas has to be taken into account. The heat transfer by conduction inside the coils and the convector has to be modelled.

Finally one can assume that the gas maintains a constant given velocity. The initial temperature distribution all over the system (coils, convectors, cover and gas) is known, as well as the schedule for firing the burners that translates into a given temperature distribution on the cover. The heating process will last until the minimum temperature in the stack has reached a prescribed value $T_m(\sim 620^\circ C)$, the burners will be firing as long as the maximum of the temperature in the stack is below a critical value $T_M(\sim 720^\circ C)$.

For a more sophisticated model we refer the reader to G.F. Harvey's paper (1977) [14]. From the physical laws of heat transfer by radiation, convection and conduction and Newton's law (for the motion of the gas) one can derive the comprehensive equation (1.1), i.e. a mathematical model for the batch annealing. At this point the mathematical problem of finding the temperature distribution $T(x,t)$ in the coil at each time t seems to be a direct problem. However, its solution (through a numerical approach) would be impossible because of the complicated geometry of the coils (thin strips rolled in about 130 windings). This is why one replaces each coil with an inhomogeneous and anisotropic solid, bounded by two coaxial cylinders of radii $r < R$ (r radius of the coil's inside surface, R that of the outside surface). Axial symmetry for the whole system (cover, gas, coils, convectors) is assumed. The conductivities of this solid are determined so that from the point of view of heat transfer (radially and vertically) it behaves the closest possible like the replaced coil. As we will show, it is in connection with the determination of the radial conductivity that the problem becomes an inverse problem. In representing the coil by a simpler geometry we first replace the windings with coaxial cylinders showing some space (gap) in between any two successive ones.

Let d_1 be the (average) thickness of the strip ($\sim 2\text{mm}$), d_2 the (average) width of the gap between the windings ($\sim .04 \times d_1$), k_s the conductivity of the steel, k_g the conductivity of the gap, we can assume with U.O. Stikker [20] additivity for the heat resistance. Then the heat resistance ρ for the heat between the two planes r_1 and r_2 (see Fig. (2.5)) is given by

$$\rho = d_1/k_s + d_2/k_g \,,$$

thus the equivalent conductivity will be

$$k_s = \frac{d_1 + d_2}{(d_1/k_s) + (d_2/k_g)}$$

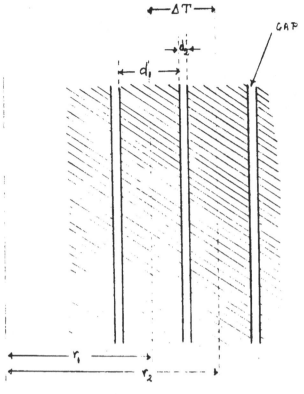

Figure 2.5

Taking into account that d_1 and d_2 depend on the temperature, U.O. Stikker (1970) [20] derived the following formula

$$k_r = \frac{1 + (d_{20}/d_{10}) + r_2\, A(\Delta T/d_{10})}{1 + (k_s/k_g)[(d_{20}/d_{10}) + r_{20}\, A((\Delta T/d_{10})]}, \qquad (2.1)$$

where d_{10} and d_{20} are the initial values of d_1 and d_2, respectively, ΔT is the jump in temperature between the two planes r_1 and r_2, while r_2 in the formula is the radius of the middle of the second winding, r_{20} is its initial value, A is a function of T. This particular parameter A is unknown since various factors can change its value. One of these factors is the history of the roll, that makes for instance the tightness of the coil random as random is the area of contact between two successive windings. For the same reason random is also the percentage of heat that flows around from each winding to the following one.

Moreover when we represent the coil as a set of cylinders we break the continuous connection of the windings. To compensate for the possible loss in the radial heat transfer when going from the real physical system to the model, one increases the radial conductivity to a value $\overline{k}_r = k_r + \eta_r$, where η_r has yet to be determined. Both A and $\overline{k}_r$ can be found, in good approximation, by using the model in the reverse direction. From experimental data one can extrapolate solutions of the (direct) mathematical problem that has been associated to the model. From these solutions one can derive the necessary information about the unknown parameters (**inverse problem**).

It should not pass unnoticed that for the control of the batch annealing, the knowledge of the temperature distribution $T(x,t)$ over the coils at each time t is not required. What one need to know, at each point in time, is its maximum and its minimum value and where they are attained. These are indeed the leading values in switching on and off the burners and to end the heating stage. For the simple model that we have described here the breakdown of equation (1.1) is as follows.

Because of the assumed axial symmetry of the system the temperature distribution is represented by a function

$$T = T(r, z, t)$$

where z is the vertical distance from the base of the stack, r is the distance from the common axis of the coils. The coordinates r and z are usually measured in centimeters, the time t in seconds, T is measured in kelvin degree.

1. Heat transfer inside each coil

$$\rho c \frac{\partial}{\partial t} T = \frac{1}{r} \frac{\partial}{\partial r} \left(r \bar{k}_r \frac{\partial T}{\partial r} \right) + k_s \frac{\partial^2 T}{\partial z^2} \tag{2.2}$$

2. On the outside surface of each coil

$$k_r \frac{\partial T}{\partial r} = \alpha_{r,0}(T_g - T) + c_{1,s}(T_c^4 - T^4) + c_{2,0}(T_g^4 - T^4) \tag{2.3}$$

where $\alpha_{r,0}$ is the convective heat transfer coefficient, $c_{1,s}$ is the radiation coefficient cover-coil, $c_{2,0}$ the radiation coefficient coil-gas

3. On the inside surface of each coil

$$-k_r \frac{\partial T}{\partial r} = \alpha_{r,i}(T_g - T) + c_{2,i}(T_g^4 - T^4) \tag{2.4}$$

where $\alpha_{r,i}$ is the convection coefficient of the inside surface, $c_{2,i}$ the radiation coefficient inside surface of the coil to the gas.

4. Bottom or upper end of each coil when convectors are present

$$\pm k_s \frac{\partial T}{\partial z} = \alpha_{z,s}(T_g - T) + G_2(T_g - T) \tag{2.5}$$

with the plus sign for the top end, negative sign for the bottom; here $\alpha_{z,s}$ is the coil surface-gas convection coefficient, G_2 is the coefficient for the heat transfer between coils via convectors. At the upper end of the top coil

$$k_s \frac{\partial T}{\partial z} = \alpha_{z,t}(T_g - T) + c_{1,t}(T_c^4 - T^4) + c_{2,t}(T_g^4 - T^4) \tag{2.6}$$

where $\alpha_{z,t}$ is the convection coefficient of the top surface of the coil, $c_{1,t}$ is the radiation coefficient cover-coil, $c_{2,t}$ the radiation coefficient coil-gas.

5. When no convector separates two coils then at their common surface the following equation has to be satisfied

$$k_{z_1} \frac{\partial T_1}{\partial z} = k_{z_2} \frac{\partial T_2}{\partial z} \tag{2.7}$$

where 1 and 2 refer to coil n.1 and coil n.2.

6. At the bottom surface of the bottom coil in absence of a convector

$$-k_{z_1}\frac{\partial T_1}{\partial z} = h(T_1 - T_2) \tag{2.8}$$

where T_2 is the temperature of the plateau, h the heat transfer coefficient plateau-coil.

For the heat transfer via the gas atmosphere a model has been chosen for the gas flow, model which is far to be refined. Yet, as one can see in more details in D. Mundie's Master Thesis (1981) [19] the results that have been obtained using the model are quite satisfactory. We assume the gas flow to be at constant uniform velocity v, and we simulate the circulation of the gas around the coils by taking into account that a certain percentage of the total mass of the gas is deflected into the convectors and forced through the central hole of stack. The following equations have been derived (see also U.O. Stikker (1970) [20]),

1. for the local gas temperature in the convector under the first coil (where the fan is located)

$$\dot{m}c_g\frac{\partial T_g}{\partial r} = \alpha_{b,z,b}(T_2 - T_g) \tag{2.9}$$

here $\dot{m}$ is the gas (total) mass flux, $\alpha_{b,z,b}$ is the coefficient of convective heat transfer gas-convector which is dependent on the gas flow velocity and the gas temperature, c_g is the specific heat of the gas.

2. for the gas temperature through the convector (if present) separating the first coil in the stack from the base plate

$$-\dot{m}_bc_g\frac{\partial T_g}{\partial r} = \alpha_{b,z,s}(T_2 - T_g) + \alpha_{z,s}(T - T_g) + G_{z,b}(T_2 - T_g) + G_{z,c}(T - T_g) \tag{2.10}$$

where $\dot{m}_b$ is the flux of that percentage of gas that flows through the spacer, while the α's are convection coefficients (base plate-gas, coil-gas) $G_{z,b}$ and $G_{z,c}$ are the heat transfer coefficients base plate to gas and gas to coil through the convector.

3. for the gas temperature between any two coils

$$-\dot{m}_{i+}c_g\frac{\partial T_g}{\partial r} = \alpha_{z,s,i}(T_i - T_g) + \alpha_{z,s,i+1}(T_{i+1} - T_g) + G_{z,i}(T_i - T_g) + G_{z,i+1}(T_{i+1} - T_g) \tag{2.11}$$

where the meaning of the parameters is clear.

4. between the outside surface of any coil and the cover

$$\dot{m}_{i-}c_g\frac{\partial T_g}{\partial z} = \alpha_{r,c}(T_c - T_g) + \alpha_{r,0,i}(T - T_g) + G_{r,\text{cov.}}(T_c - T_g) + C_{r,i,\text{coil}}(T - T_g) \tag{2.12}$$

where the parameters again have clear meaning.

5. between the top of the top coil and the cover

$$-\dot{m}_tc_g\frac{\partial T_g}{\partial r} = \alpha_{z,c}(T_c - T_g) + \alpha_{z,t}(T - T_g) + G_{z,c}(T_c - T_g) + G_{z,t}(T - T_g) \tag{2.13}$$

where again the meaning of the parameters and the unknown temperatures is clear.

6. finally for the gas temperature distribution in its (downward) vertical flow in the central hole of the stack

$$-\dot{\overline{m}}_{i+}c_g\frac{\partial T_g}{\partial z} = \alpha_{r,in,i}(T - T_g)G_{r,i,\text{coil}}(T - T_g) \tag{2.14}$$

where $\dot{\overline{m}}_{1+}$ is the percentage of the gas flux along the inside surface of the i-th coil, $\alpha_{r,in,i}$ is the convection coefficient in the gas contact with that surface.

The mathematical model, within a reasonable approximation, is now complete. Note, however, that most of the physical parameters can be changed and used as control parameters. In particular: (i) the gas atmosphere fluxes (by means of deflectors installed on the inside surface of the cover, (ii) the gas velocity, that effects the value of all the $\dot{m}$'s, (iii) the coefficient of convective heat transfer through the convectors, by changing their design.

The mathematical problem associated with this model is highly non-linear, because of the temperature dependence of the conductivities k_r and k_z. To make the problem even more difficult (see Stikker's formula), k_r depends also on the radial derivative of the temperature. Yet, solutions found by a numerical (computer) approach using a finite difference method were in satisfactory agreement with the experimental data (that have been provided by a steel company). We used the Peaceman-Rachford method (backward-forward finite differences). Accordingly, equation (2.2) has been replaced with the following system of equations at each node of a non-uniform mesh

$$\rho c\left(T(r,z,t-\Delta t)\right)\left[\frac{T(r,z,t)-T(r,z,t-\Delta t)}{\Delta t}\right]$$

$$= \frac{2}{\Delta r_1 + \Delta r_2}\left\{\overline{k}_r\left(r-\tfrac{1}{2}\Delta r_1, T(r-\tfrac{1}{2}\Delta r_1,z,t-\Delta t)\right)\left[\frac{T(r-\Delta r_1,z,t)-T(r,z,t)}{\Delta r_1}\right]\right.$$

$$\left. +\overline{k}_r\left(r+\tfrac{1}{2}\Delta r_2, T(r+\tfrac{1}{2}\Delta_2,z,t\Delta t)\right)\left[\frac{T(r+\Delta r_2,z,t)-T(r,z,t)}{\Delta r_2}\right]\right\}$$

$$+\frac{1}{2r}\left\{\overline{k}_r\left(r-\tfrac{1}{2}\Delta r_1, T(r-\tfrac{1}{2}\Delta r_1,z,t-\Delta t)\right)\left[\frac{T(r,z,t)-T(r-\Delta r_1,z,t)}{\Delta r_1}\right]\right. \tag{2.15}$$

$$\left. +\overline{k}_r\left(r+\tfrac{1}{2}\Delta r_2, T(r+\tfrac{1}{2}\Delta r_2,z,t-\Delta t)\right)\left[\frac{T(r+\Delta r_2,z,t)-T(r,z,t)}{\Delta r_2}\right]\right\}$$

$$+\frac{2}{\Delta z_1 + \Delta z_2}\left\{k_z\left(T(r,z-\tfrac{1}{2}\Delta z_1,t-\Delta t)\right)\left[\frac{T(r,z-\Delta z_1,t-\Delta t)-T(r,z,t-\Delta t)}{\Delta z_1}\right]\right.$$

$$\left. +k_z\left(T(r,z,+\tfrac{1}{2}\Delta z_2,t-\Delta t)\right)\left[\frac{T(r,z+\Delta z_2,t-\Delta t)-T(r,z,t-\Delta t)}{\Delta z_2}\right]\right\}$$

$$\rho c\Big(T(r,z,t-\Delta t)\Big)\left[\frac{T(r,z,t)-T(r,z,t-\Delta t)}{\Delta t}\right]$$

$$=\frac{2}{\Delta r_1+\Delta r_2}\left\{\overline{k}_r\Big(r-\tfrac{1}{2}\Delta r_1,T(r-\tfrac{1}{2}\Delta r_1,z,t-\Delta t)\Big)\left[\frac{T(r-\Delta r_1,z,t-\Delta t)-T(r,z,t-\Delta t)}{\Delta r_1}\right]\right.$$

$$\left.+\overline{k}_r\Big(r+\tfrac{1}{2}\Delta r_2,T(r+\tfrac{1}{2}\Delta r_2,z,t-\Delta t)\Big)\left[\frac{T(r+\Delta r_2,z,t-\Delta t)-T(r,z,t-\Delta t)}{\Delta r_2}\right]\right\}$$

$$+\frac{1}{2r}\left\{\overline{k}_r\Big(r-\tfrac{1}{2}\Delta r_1,T(r-\tfrac{1}{2}\Delta r_1,z,t-\Delta t)\Big)\left[\frac{T(r,z,t-\Delta t)-T(r-\Delta r_1,z,t-\Delta t)}{\Delta r_1}\right]\right.\qquad(2.16)$$

$$\left.+\overline{k}_r\Big(r+\tfrac{1}{2}\Delta r_2,T(r+\tfrac{1}{2}\Delta r_2,z,t-\Delta t)\Big)\left[\frac{T(r+\Delta r_2,z,t-\Delta t)-T(r,z,t-\Delta t)}{\Delta r_2}\right]\right\}$$

$$+\frac{2}{\Delta z_1+\Delta z_2}\left\{k_z\Big(T(r,z-\Delta z_1,t-\Delta t)\Big)\left[\frac{T(r,z-\Delta z_1,t)-T(r,z,t)}{\Delta z_1}\right]\right.$$

$$\left.+k_z\Big(T(r,z+\Delta z_2,t-\Delta t)\Big)\left[\frac{T(r,z+\Delta z_2,t)-T(r,z,t)}{\Delta z_2}\right]\right.$$

Summarizing: we first determine the parameter k and those convection coefficients that simulate complicated physical structures, by using given (experimental) solution of the mathematical problem (inverse problem). Then (in a routine manner) we use the model to design a more efficient (time-energy) procedure and technology, by solving (numerically) the (direct) mathematical problem with the following input data,

1 - number of coils in each of the four stacks,

2 - heights of coils, base plate and eventual convectors,

3 - outer radius of coils,

4 - inner radius of coils,

5 - outer radius of base plate,

6 - thickness of steel sheets in each coil,

7 - minimum number of grid points,

8 - fan rate,

9 - percentages of gas flow through the convectors,

10 - initial temperature in the coils,

11 - initial gas temperature,

12 - heat convection coefficients of each convector,

13 - physical parameters for heat transfer throughout the system,

14 - value of the final minimum temperature T_m,

15 - critical temperature T_M.

To conclude: modelling the batch annealing leads to the solution of an inverse problem, where the structure of A and f in equation (1.1) is known, but to explicit them ne parameters have to be determined.

3. CLASSICAL INVERSE PROBLEMS IN CLASSICAL ONE PARTICLE DYNAMICS

Classical dynamics provides clear examples of different kinds of inverse problems. Whereas the **direct** problem is formulated as follows,

given the force field $F(x, \dot{x}, t)$, the mass m, the initial position $x(0) = x_0$ and velocity $\dot{x}(0) = x_0$ of the particle, find the function $x(t)$, $t \geq 0$, (the motion of the particle) such that

$$m\ddot{x}(t) = F(x(t), \dot{x}(t), t), \quad t > 0 \tag{3.1}$$

$$x(0) = x_0$$

$$\dot{x}(0) = \dot{x}_0 \,,$$

in other words equation (1.1) being fully determined, find its solution.

An inverse problem is formulated as follows,

given some properties of the motion, find the force field and/or all compatible initial conditions, that is given some information about the solution ϕ fine in (1.1) A and/or f.

We will report here some classical examples of the three most relevant types of inverse problems in classical mechanics; for further examples and formal methods for solution we refer the reader to A.S. Galiullin (1984) [8].

3a – Newton's problem: In this problem we are looking for the force field F such that Kepler's laws are satisfied, that is

1 - all trajectories be ellipses with one of the two foci in common

2 - the sector (or areal) velocity with respect to that focus 0_1 be constant

3 - the square of the time taken by the particle to complete a revolution around the said focus be proportional to the cube of the major semiaxis a of the orbit.

If e is the eccentricity of the orbit, 0_2 the second focus, (x, θ) the polar coordinate of the moving particle with pole 0_1, θ being the angle between the vector $0_2 0_1$ and the (position) vector x, the equation of a trajectory on any given plane that contains 0_1 and 0_2 is

$$x = \frac{p}{1 + e \cos \theta}$$

where p is a constant.

For the sector velocity $\dot{A}$ we have

$$\dot{A} = \frac{1}{2} \mid x \mid^2 \dot{\theta} = c, \tag{3.2}$$

where c is a second arbitrary constant. $| x |$ denotes the length of the position vector of the particle. In a cartesian (orthogonal) coordinate system in the plane of the orbit, with origin 0_1, x_1-axis the line $0_2 0_1$ and x_2-axis the perpendicular to $0_2 0_1$ (through 0_1), let x_1 and x_2 be also the cartesian components of the position vector x, so that $| x | = \sqrt{x_1^2 + x_2^2}$, and

$$\dot{A} = \frac{1}{2}(x_1 \dot{x}_2 - x_2 \dot{x}_1) = c. \tag{3.3}$$

By differentiating both sides of this equation we get

$$x_1 \ddot{x}_2 - x_2 \ddot{x}_1 = 0,$$

and with $F = (X_1, X_2, X_3)$ where X_i are the cartesian components of F, by equation (3.1) (Newton's law) we have

$$\begin{aligned} x_1 X_2 - x_2 X_1 &= 0, \\ X_3 &= 0. \end{aligned} \tag{3.4}$$

Equations (3.4) imply that the force field F is at each point parallel to the position vector x and therefore it is a central field with centre 0_1. Moreover, by the conservation of the sector velocity (equation (3.3)) and Binet's formula for the component $\ddot{x}_r$ of the acceleration $\ddot{x}$ in the direction of the position vector x, we have

$$\ddot{x}_r := \ddot{x} \cdot x / | x | = -\frac{4c^2}{| x |^2} \left[\frac{d^2}{d\theta^2} \left(\frac{1}{| x |} \right) + \frac{1}{| x |} \right] \tag{3.5}$$

Then taking into account the given equation of the orbit we have

$$\ddot{x}_r = -4c^2 / (p \, | x |^2),$$

and by Newton's law

$$F = -4c^2 m \, x / (p \, | x |^3).$$

Denote with b the length of the second semi-axis of the orbit. In a complete revolution around 0_1, the vector position of the particle covers an area equal to $\pi a b$. Hence the time T to complete such a revolution is

$$T = \pi a b / c,$$

then, since $p = a(1 - e^2) = b^2 / a$,

$$T^2 = \pi^2 a^2 b^2 / c^2 = (\pi^2 p / c^2) a^3,$$

which represents the third Kepler's law.

3b – Bertrand's problem: It is slightly more general than Newton's problem, in the sense that less information about the functions A and f in equation (1.1) is given. The problem is formulated as follows, given that

i) all trajectories lie on a plane (x_1, x_2),

ii) each of them is a conic section, with a common focus

iii) no restriction is imposed by the force field on the initial position and velocity of the particle,

find the force field $F(x_1, x_2) = [X_1(x_1, x_2), X_2(x_1, x_2)]$.

With an appropriate choice of the coordinate system in the plane of the conic section, its equation can be written in the form

$$\sqrt{x_1^2 + x_2^2} = -e x_1 + p \tag{3.6}$$

where e and p are non-negative constants.

By differentiating twice both sides of (3.6), when possible $(|x| \neq 0!)$, we obtain

$$\frac{x_1 \dot{x}_1 + x_2 \dot{x}_2}{\sqrt{x_1^2 + x_2^2}} = \frac{x_1 \dot{x}_1 + x_2 \dot{x}_2}{|x|} = -e \dot{x}_1$$

$$-\frac{1}{|x|^3}(x_1 \dot{x}_1 + x_2 \dot{x}_2)^2 + \frac{\dot{x}_1^2 + \dot{x}_2^2}{|x|} + \frac{x_1}{|x|}\ddot{x}_1 + \frac{x_2}{|x|}\ddot{x}_2 = -e \ddot{x}_1 \tag{3.7}$$

$$\frac{1}{|x|^3}(x_1 \dot{x}_2 - x_2 \dot{x}_1)^2 + \frac{x_1}{|x|}\ddot{x}_1 + \frac{x_2}{|x|}\ddot{x}_2 = -e \ddot{x}_1$$

and by recourse to Newton's law

$$\frac{1}{|x|^3}(x_1 \dot{x}_2 - x_2 \dot{x}_1)^2 + \frac{x_1}{|x|}\frac{X_1}{m} + \frac{x_2}{|x|}\frac{X_2}{m} = -e\frac{X_1}{m} \tag{3.8}$$

Choose the initial data for $x_1(0)$, $x_2(0)$, $\dot{x}_1(0)$ and $\dot{x}_2(0)$ so that $x_1(0)\dot{x}_2(0) = x_2(0)\dot{x}_1(0)$.

By expressing e in terms of $x_1, x_2, \dot{x}_1, \dot{x}_2$ through the first of equations (3.7) we can rewrite (3.8) in the following form

$$\frac{m}{|x|^2}(x_1 \dot{x}_2 - x_2 \dot{x}_1)^2 = \frac{x_2}{x_1}(X_1 \dot{x}_2 - X_2 \dot{x}_1),$$

then if $x_2(0) \neq 0$, we have

$$X_1 \dot{x}_2(0) - X_2 \dot{x}_1(0) = \frac{\dot{x}_1(0)}{x_1(0)}(X_1 x_2(0) - X_2 x_1(0)) = 0$$

and consequently by the arbitrariness of $(x_1(0), x_2(0))$ and of the choice the axes of the coordinate system, the force $F(x_1, x_2)$ is at each point (x_1, x_2) parallel to the line joining that point with the origin (the common focus of the conic sections).

Thus there exists a function $V = V(x_1, x_2)$, defined everywhere except at the origin $(0, 0)$, such that

$$X_1 = V x_1, \qquad X_2 = V x_2.$$

As a consequence of Newton's law we have

$$x_1 \ddot{x}_2 - x_2 \ddot{x}_1 = \frac{d}{dt}(x_1 \dot{x}_2 - x_2 \dot{x}_1) = 0$$

hence the sector velocity $\dot{A} = \frac{1}{2}(x_1\dot{x}_2 - x_2\dot{x}_1)$ is constant, say c.

Equation (3.8) then yields

$$\frac{4c^2}{\mid x \mid^3} = -ex_1\frac{V}{m} - \frac{x_1^2}{\mid x \mid}\frac{V}{m} - \frac{x_2^2}{\mid x \mid}\frac{V}{m} = -ex_1\frac{V}{m} - \mid x \mid\frac{V}{m}$$

and since

$$\mid x \mid = -ex_1 + p$$

we have

$$V = -\frac{4c^2 m}{p\mid x \mid^3}$$

Hence F is a gravitational field with components

$$X_1 = -\frac{4c^2 m}{p\mid x \mid^2}\frac{x_1}{\mid x \mid}, \qquad X_2 = -\frac{4c^2 m}{p\mid x \mid^2}\frac{x_2}{\mid x \mid}.$$

If in Bertrand's assumption we replace assumption (ii) with

ii*) all trajectories are conic section with (common) **centre**,

with a proper choice of the coordinate system their equation can be written as follows

$$Ax_1^2 + Bx_2^2 = 1. \tag{3.9}$$

By differentiating this equation twice, we have

$$Ax_1\dot{x}_1 + Bx_2\dot{x}_2 = 0 \tag{3.10}$$

$$A\dot{x}_1^2 + B\dot{x}_2^2 + Ax_1\ddot{x}_1 + Bx_2\ddot{x}_2 = 0$$

and with recourse to Newton's law

$$A\dot{x}_1^2 + B\dot{x}_2^2 + Ax_1\frac{X_1}{m} + Bx_2\frac{X_2}{m} = 0 \tag{3.11}$$

Obvious manipulations lead to

$$-B\frac{x_2\dot{x}_1\dot{x}_2}{x_1} + Bx_2^2 - B\frac{x_2\dot{x}_2}{\dot{x}_1}\frac{X_1}{m} - Bx_2\frac{X_2}{m} = 0$$

that is

$$\frac{\dot{x}_2}{x_1}(x_1\dot{x}_2 - x_2\dot{x}_1) + \frac{x_2}{\dot{x}_1}\left(\dot{x}_1\frac{X_2}{m} - \dot{x}_2\frac{X_1}{m}\right) = 0,$$

which is independent on the orbit parameters A and B and therefore it just represents a property of the force field. Since the field does not impose any restriction on the initial values of the position vector x and the velocity $\dot{x}$, we may choose such values, i.e. $x_1(0)$, $x_2(0)$, $\dot{x}_1(0)$ and $\dot{x}_2(0)$, so that

$$x_1(0)\dot{x}_2(0) - x_2(0)\dot{x}_1(0) = 0.$$

This yields again (if $x_2(0) \neq 0$), with $F(0) = F(x_1(0), x_2(0))$,

$$\dot{x}_1(0)X_2(0) - \dot{x}_2(0)X_1(0) = 0.$$

By the arbitrariness in the choice of the initial position and orbit, we have that the force field is central, and there exists a function $V(x_1, x_2)$ such that

$$X_1 = V x_1, \qquad X_2 = V x_2.$$

By replacing X_1 and X_2 into equation (3.11) we obtain

$$A\dot{x}_1^2 + B\dot{x}_2^2 + (Ax_1^2 + Bx_2^2)\frac{V}{m} = A\dot{x}_1^2 + B\dot{x}_2^2 + \frac{V}{m} = 0,$$

then

$$\frac{V}{m} = -A\dot{x}_1^2 - B\dot{x}_2^2$$

which yields

$$\frac{\dot{V}}{m} = -2A\dot{x}_1\ddot{x}_1 - 2B\dot{x}_2\ddot{x}_2 = -2(Ax_1\dot{x}_1 + Bx_2\dot{x}_2)\frac{V}{m} = 0$$

thus $V(t) = V(x_1(t), x_2(t))$ is constant $(= V_0)$ in the motion of our particle, on the curve (3.9), and we can write

$$X_1 = V_0 x_1, \qquad X_2 = V_0 x_2 \,,.$$

Both Newton's problem and Bertrand's problem are examples of inverse problems where solution can be found and, up to constants, such solution is unique. We are not that lucky with the following inverse problem which is similar to example 1 in paragraph 2.

3c - Suslov's problem (1890) [21]:   The information we have about the force field (about equation (1.1)) is now

i) the force field, on the plane (x_1, x_2), is conservative with unknown potential $U(x_1, x_2)$,

ii) one trajectory of the particle is given, in the form

$$\omega(x_1, x_2) = 0.$$

This means that with proper initial conditions $x_1(0) = x_{10}$, $x_2(0) = x_{20}$, $\dot{x}_1(0) = \dot{x}_{10}$, $\dot{x}_2(0) = \dot{x}_{20}$, we have

$$\omega(x_1(t), x_2(t)) = 0 \quad \text{for all} \quad t \geq 0, \tag{3.12}$$

and with $T = m\,|\,\dot{x}\,|^2/2$, the kinetic energy of the particle,

$$T - U = \text{constant} \quad = h.$$

Differentiating (3.12) w.r. to time t, we obtain

$$\frac{\partial \omega}{\partial x_1}\dot{x}_1 + \frac{\partial \omega}{\partial x_2}\dot{x}_2 = \text{grad}\,\omega \cdot \dot{x} = 0$$

in the motion of the particle on the curve $\gamma := \{(x_1, x_2) : \omega(x_1, x_2) = 0\}$. Then we can write

$$\dot{x}_1 = M(x_1, x_2)\frac{\partial \omega}{\partial x_2} + \phi_1(\omega, x_1, x_2)$$

$$\dot{x}_2 = -M(x_1, x_2)\frac{\partial \omega}{\partial x_1} + \phi_2(\omega, x_1, x_2),$$

with $\phi_1(0, x_1, x_2) = \phi_2(0, x_1, x_2) = 0$, i.e. with $\phi_1 = \phi_2 = 0$ on γ. Each choice of the function M will define the potential U (i.e. the force field) through the equation (energy conservation)

$$U = T - h = \frac{1}{2}m\left[\left(\frac{\partial \omega}{\partial x_1}\right)^2 + \left(\frac{\partial \omega}{\partial x_2}\right)^2\right]M^2(x_1, x_2) + \Phi(\omega, x_1, x_2) - h,$$

with $\Phi(0, x_1, x_2) = 0$ for all x_1, x_2, otherwise arbitrary.

Clearly Suslov's problem is largely undetermined, showing infinitely many solutions.

Bertrand's problem can now be restated, by further reducing the information, namely by eliminating (i), replacing (ii) with

ii*) the trajectory of the particle is the conic section

$$\omega(x_1, x_2) = \sqrt{x_1^2 + x_2^2} + ex_1 - p = 0, \tag{3.13}$$

and (iii) with

iii**) the unknown force field is conservative.

Then the potential of the force field has the form

$$\begin{aligned}
U(x_1, x_2) &= \frac{m}{2}M^2(x_1, x_2)\left[\left(\frac{x_1}{|x|} + e\right)^2 + \left(\frac{x_2}{|x|}\right)^2\right] + \Phi(\omega, x_1, x_2) - h \\
&= \frac{m}{2}M^2(x_1, x_2)\left[\frac{x_1^2 + x_2^2}{|x|^2} + \frac{2x_1 e}{|x|} + e^2\right] + \Phi(\omega, x_1, x_2) - h \\
&= \frac{m}{2}M^2(x_1, x_2)\left[1 + \frac{2x_1 e}{|x|} + e^2\right] + \Phi(\omega, x_1, x_2) - h,
\end{aligned}$$

and since $\omega = 0$ implies $|x| + ex_1 = p$ and $\Phi(0, x_1, x_2) = 0$, on the given trajectory γ we have

$$\begin{aligned}
U(x_1, x_2) &= \frac{m}{2}M^2\left[\frac{p}{|x|} + e\left(\frac{x_1}{|x|} + e\right)\right] - h = \frac{m}{2}M^2\left[2\frac{p}{|x|} - 1 + e^2\right] - h \\
&= \overline{M}^2\frac{p}{|x|} - \overline{h}.
\end{aligned}$$

If we assume the (still unknown) force field central, with centre the centre of the conic section, we would have $x_1\dot{x}_2 - x_2\dot{x}_1 = \text{constant} =: 2c$, and $\overline{M}^2 = 4c^2/p^2$,

$$U = 4\frac{c^2}{r}\frac{1}{|x|} + \text{const.} \quad \text{on} \quad \gamma.$$

3d - Meshcherskii's problem (1949) [18]: It deals with "modelling" the motion of a particle with variable mass $m = m(t)$ in a vertical plane (x_1, x_2), with the x_2-axis vertical (upward). The motion of the particle is given, as well as the force field (gravity, drag).

With $x_1 = \phi_1(t)$, $x_2 = \phi_2(t)$, given, by Newton's law we have

$$m(t)\ddot{\phi}_1 = \dot{m}(t)(\mu_1 - 1)\dot{\phi}_1 - m(t) f(x_2, |\dot{x}|)\dot{\phi}_1/|\dot{x}| \qquad (3.14)$$

$$M(t)\ddot{\phi}_2 = \dot{m}(t)(\mu_2 - 1)\dot{\phi}_2 - m(t) f(x_2, |\dot{x}|)\dot{\phi}_2/|\dot{x}| - mg,$$

where μ_i is the ratio between the i-th component of the velocity of the escaping mass and $\dot{x}_i$, $i = 1, 2,$, $|\dot{x}| = \sqrt{\dot{\phi}_1^2 + \dot{\phi}_2^2}$.

The problem is: Find the flow rate of the escaping mass $\dot{m}$, the parameters μ_i, $i = 1, 2,$, so that the required motion of the rocket ("the particle") is achieved.

It is a control problem, the controls being the functions $\dot{m}(t)$, $\mu_1(t)$, $u_2(t)$. Here the structure of the functions A, f in equation (1.1) is known, the inverse problem requires to find the controls given the solution ϕ of (1.1) (note that (1.1) is comprehensive of the initial conditions). By plugging in (3.14) the given solution $x_1 = \phi_1(t)$, $x_2 = \phi_2(t)$ and solving for the controls we get

$$\mu_1(t) = 1 + \frac{m}{\dot{m}}\left[\frac{\ddot{\phi}_1}{\dot{\phi}_1} + \frac{f(\phi_2(T), v)}{v}\right]$$

$$\mu_2(t) = 1 + \frac{m}{\dot{m}}\left[\frac{\ddot{\phi}_2}{\dot{\phi}_2} + \frac{g}{\dot{\phi}_2} + \frac{f(\phi_2(t), v)}{v}\right], \quad v = \sqrt{\dot{\phi}_1^2 + \dot{\phi}_2^2} = |\dot{x}|.$$

Two equations for the three unknown controls. To "close" the system one can add either the knowledge of the relative speed u of the escaping mass, i.e.

$$u = \sqrt{(\mu_1 - 1)^2\dot{\phi}_1^2 + (\mu_2 - 1)^2\dot{\phi}_2^2},$$

or the angle

$$\alpha = \arccos\frac{(\mu_1 - 1)\dot{\phi}_1^2 + (\mu_2 - 1)\dot{\phi}_2^2}{uv}$$

between the velocity of the escaping mass and that of the rocket.

4. THREE GENERAL FORMULATIONS FOR INVERSE PROBLEMS IN DYNAMICS

The most common inverse problems in dynamics fall in one of the following three types:

1. *The fundamental problem* With $x = (x_1, x_2, \ldots, x_n)$, $\dot{x} = (\dot{x}_1, \dot{x}_2, \ldots, \dot{x}_n)$, $\ddot{x} = (\ddot{x}_1, \ddot{x}_2, \ldots, \ddot{x}_n)$, construct the system

$$\ddot{x} = F(x, \dot{x}, t),$$

i.e. find the vector $F = (F_1, F_2, \ldots, F_n)$, such that

$$\Omega : \omega_j(x, \dot{x}, t) = c_j \qquad j = 1, 2, \ldots m \leq 2n$$

is an integral manifold.

2. *The restoration problem*   Given the system

$$\ddot{x} = F_0(x, \dot{x}, v, t)$$

find (the control)

$$v = (v_1, v_2, \ldots, v_k),$$

with $v_j = v_j(x, \dot{x}, t)$, $j = 1, 2, \ldots, k$,   so that

$$\Omega : \omega_h(x, \dot{x}, t) = c_h \qquad h = 1, 2, \ldots, m \leq 2n$$

is an integral manifold.

3. *The closure problem*   It means that we have to "close" the equation of motion that otherwise would be "open". More precisely, given the system

$$\ddot{x} = F_{0,c}(x, x, u, u, t)$$

find a (closing) system of equations

$$\ddot{u}_h = U_h(x, \dot{x}, u, \dot{u}, t) \qquad h = 1, 2, \ldots, r,$$

for the unknown function $u = (u_1, u_2, \ldots, u_r)$,   so that

$$\Omega : \omega_j(x, \dot{x}, t) = c_j \qquad j = 1, 2, \ldots, m,$$

is an integral manifold.

General approaches for a formal solution of the above problems can be found in A.S. Galiullin's book (see references).

Newton's and Bertrand's problems are examples of inverse problems type 1. Meshcherskii's problem is an example of inverse problem of type 2. Dynamical systems with a second mechanical system as a servo-mechanism will provide examples of inverse problems of type 3.

Note that the integral manifolds $\omega_j(x, \dot{x}, t) = c_j$ represent an information about the system (A, f). As one can understand from the practical examples of paragraph 2, this form of information is purely theoretical. It would be more realistic to assume the c_j's fixed constants (usually zero); this is indeed the case for instance, when a certain finite set of solutions of the dynamical system is given.

If we extend the above general inverse problems so to include the case where all or part of the constants c_j's are fixed, then Suslov's problem becomes an example of inverse problem of type 1, and the practical cases just mentioned would be included (see in particular example 1 of paragraph 2). However, it should be noticed (see Suslov's problem) that in general the knowledge of few solutions may not be of great help in the determination of the dynamical system and consequently in finding solutions that correspond to different initial conditions. We believe that this problem is important enough to draw the attention of the researchers in this field. We also feel that it should be approached with the setting of general dynamical systems, where better insight can be gained by recourse to a variety of applied inverse problems. For this reason, next paragraph will be devoted to an introduction on general dynamical systems, while the following paragraph 5 deals with a "visual" inverse problem: image representation by approximate solutions of an inverse problem (in "general" dynamical systems).

4 - General dynamical systems: To investigate the qualitative behaviour of the solutions of an autonomous system of ordinary differential equations (see G. Birkhoff (1927)[7]) independently from any underlying analytical representation, the following abstract setting has been introduced.

Let (X, d) be a complete, compact metric space and let T be either the set $\mathbf{R}^+$ of non-negative real numbers with the usual metric topology or the set $\mathbf{Z}^+$ of non-negative integers with the discrete metric topology. A (crisp) semi-dynamical system on the **state space** X is the quadruple (X, d, T, w) where w is a single-valued mapping satisfying the four axioms

(1) w is defined for all (x, t) in $X \times T$,

(2) $w(x, 0) = x$ for all $x \in X$,

(3) $w(x, s + t) = w(w(x, s), t)$, for all $x \in X$ and $s, t, \in T$,

(4) w is continuous in t for each fixed x and upper semi-continuous in x for each fixed t.

A generalized semi-dynamical system is obtained by allowing the mapping w to be multivalued, i.e. the image of a point x in X is a subset of X with cardinality greater than or equal to 1. With P.E. Kloeden (1974) [16] for each $(x, t) \in X \times T$ the attainable set $w(x, t)$ is the set of all points in X that are reached at time t from the initial point x. The axioms that define a generalized semi-dynamical system are as follows,

(1') w is defined for all $(x, t) \in X \times T$,

(2') $w(x, 0) = \{x\}$ i.e. is the singleton x in X, for all $x \in X$,

(3') $w(x, s + t) =: w(w(x, s), t) := \{w(y, t) : y \in w(x, s)\}$ for all $x \in X$, $s, t \in T$,

(4') w is continuous in t for each fixed x, upper semi-continuous in x for each fixed t.

For each point $x \in X$, a trajectory in a generalized semi-dynamical system is defined as a single-valued mapping $\phi(x_0, .) : T \to X$, such that

$$\Phi(x_0, t) \in w(x_0, t) \quad \text{for all} \quad t \in T.$$

The above generalization could be further weakened by replacing the map w of X into the set $\mathcal{P}$ of its parts, with a map $W : \mathcal{P} \to \mathcal{P}$. Clearly, for each $t \in T$ and $w(., t)$, a map $W(., t) : \mathcal{P} \to \mathcal{P}$ is uniquely defined, by setting $W(A, t) := \cup_x \{w(x, t) : x \in A\}$. Yet for our application to image representation by a dynamical system we need to consider an even more general setting. The setting will be that of fuzzy dynamical systems (see P.E. Kloeden (1982)[16]).

To be more specific, let $I = [0, 1]$, following L.A. Zadeh (1965) [22], a fuzzy subset of X is a mapping $U : X \to I$. For each $x \in X$, the numerical value $u(x)$ represents the degree of membership of the point x in the subset u, hence $u(x) = 0$ is the non-membership, $u(x) = 1$ is full membership, and $0 < u(x) < 1$ represents partial membership. The support of the fuzzy set u is the set all $x \in X$ with partial or full membership in u, we denote this set by $\text{supp}\, u$, thus $\text{supp}\, u := \{x \in X : u(x) > 0\}$. If A, is a nonempty subset of X, its characteristic function

$$X_A(x) = 1 \qquad \text{if} \quad x \in A$$

$$X_A(x) = 0 \qquad \text{otherwise},$$

is a fuzzy subset of X; such fuzzy sets will be considered as the singleton fuzzy subsets of X while the empty fuzzy subset u^0 of X is the zero mapping $u^0(x) = 0$ for all $x \in X$.

Given the compact metric space (X, d), $\mathcal{F}(X)$ will denote the class of all fuzzy subsets of X. We shall consider the subclass $\mathcal{F}*(X)$ of $\mathcal{F}(X)$ of all the fuzzy subsets of X that are

1 - upper semi-continuous (u.s.c) on (X, d),

2 - **normal**, that is there exists x_0 in X such that $u(x_0) = 1$.

For each $0 < \alpha \leq 1$, denote with $[u]^\alpha$ the α-level set of u, i.e. $[u]^\alpha := \{x \in X : u(x) \geq \alpha\}$, and with $[u]^0$ the closure of $\text{supp}\, u$.

It is easy to recognize that for each $u \in \mathcal{F}*(X)$ all sets $[u]^\alpha$, $0 \leq \alpha \leq 1$, are nonempty compact subsets of X.

Now consider the set $\mathcal{K}(X)$ of all nonempty closed subsets of X (which includes X and its singletons $\{x\}$). Let h be the Hausdorff distance function on $\mathcal{K}(X)$, i.e. for each $A, B \in \mathcal{K}(X)$,

$$h(A, B) := \max \{D(A, B), D(B, A)\}, \tag{4.1}$$

where

$$D(A, B) := \operatorname*{supremum}_{x \in A} \left(\operatorname*{infimum}_{y \in B} d(x, y) \right).$$

$(\mathcal{K}(X), h)$ is a compact metric space and contains in particular all the level sets $[u]^\alpha$ of each fuzzy set $u \in \mathcal{F}*(X)$. Following P. Diamond and P.E. Kloeden (1990) [23], a metric d_∞ on $\mathcal{F}*(X)$ will be defined as follows

$$d_\infty(u, v) = \operatorname*{supremum}_{0 \leq \alpha \leq 1} \{h([u]^\alpha \cdot [v]^\alpha)\} \quad \text{for all} \quad u, v, \in \mathcal{F}*(X).$$

The metric space $(\mathcal{F}*(X), d_\infty)$ is complete.

A fuzzy semi-dynamical system on $\mathcal{F}*(X)$ is a quadruple $(\mathcal{F}*(X), d_\infty, T, \tilde{W})$ where $\tilde{W}$ is a single valued mapping $\mathcal{F}*(X) \to \mathcal{F}*(X)$ satisfying the four axioms

1") $\tilde{W}$ is defined for all (u, t) in $\mathcal{F}*(X) \times T$,

2") $\tilde{W}(u, 0) = u$ for all $u \in \mathcal{F}*(X)$,

3") $\tilde{W}(u, s+t) = \tilde{W}(\tilde{W}(u, s), t)$, for all $u \in \mathcal{F} * (X)$ and $s, t \in T$,

4") $\tilde{W}$ is continuous in t for each fixed u and upper semi-continuous in u for each fixed t.

One can recognize in the above definition a slight generalization of the definition in Kloeden (1982) [16], where axiom (2") is replaced by: $\tilde{W}(u, 0) = [u]^0$. We wanted here the case when uncertainty is also about the initial state of the system be included. Notice that this setting for a semi-dynamical system encompasses the previously reported kinds of semi-dynamical systems. The semi-dynamical system where $W : \mathcal{K}(X) \to \mathcal{K}(X)$, is for example the restriction of a fuzzy semi-dynamical system to the subclass of $\mathcal{F} * (X)$ which consists of the characteristic functions of all sets in $\mathcal{K}(X)$. In particular it includes the classical deterministic dynamical systems, i.e. those defined by the solution operator $w(x, t)$ of a system of autonomous ordinary differential equations

$$\dot{x} = X(x), \quad x = (x_1, x_2, \ldots, x_n),$$

with unique solution, and their discrete version

$$x(n) = X(x(n-1)), \qquad n = 1, 2, \ldots.$$

It includes as well the dynamical systems represented by a system of autonomous ordinary differential with non-unique solution for the initial value problem. Very little has to be changed to extend the setting to general function spaces. We leave it to the imagination of the reader.

The formulation of an inverse problem for a general dynamical system will be worded as follows: given some features of some trajectories find the mapping W in a certain given class $\mathcal{D}$ of dynamical systems. This problem is usually "improperly" posed, in the sense that, with M.M. Lavrentiev (1967) 17], either

a) the solution does not exist,
   and/or

b) it is not unique,
   and/or

c) it is not sufficiently "regular".

However, one is usually satisfied with optimal and/or approximate solutions.

In view of the application to image representation, in the sequel we shall assume $T = \mathbf{Z}^+$.

Moreover, the kind of semi-dynamical systems we shall be considering is such that one can extend them to a dynamical system with $W : \mathcal{K}(X) \times \mathbf{Z} \to \mathcal{K}(X)$, by defining for each $t > 0$ and $B \in \mathcal{K}(X)$,

$$W(B, -t) = C \quad \text{such that} \quad W(C, t) = B, \tag{4.2}$$

whenever such C exists in $\mathcal{K}(X)$.

One of the given features in the inverse problem for a dynamical system can be represented by the "attractors". A set $A \in \mathcal{K}(X)$ is an attractor for a family $\mathcal{G}(X)$ of sets in $\mathcal{K}(X)$ if and only if

$$\lim_{t \to \infty} h(W(S,t), A) = 0 \quad \text{for all} \quad S \in \mathcal{G}(X). \tag{4.3}$$

Similarly a fuzzy set $u* \in \mathcal{F}*(X)$ is an attractor for a given subclass $\Phi(X)$ of fuzzy sets in $\mathcal{F}*(X)$ if and only if

$$\lim_{t \to \infty} h_\infty(W(v,t), u*) = 0 \quad \text{for all} \quad v \in \Phi(X). \tag{4.4}$$

When $\mathcal{G}(X) = \mathcal{K}(X)$ (or $\Phi(X) = \mathcal{F}(*X)$) we shall call the (unique) attractor $A(u*)$ the **attractor** of the semi-dynamical system (the fuzzy semi-dynamical system).

As we shall see in the following section, image representation leads to an inverse problem of the kind: given the attractor $(u*)$ A and a class $\mathcal{D}$ of semi-dynamical systems, find in $\mathcal{D}$ the (fuzzy) semi-dynamical system.

5 - Image representation and dynamical systems With M.F. Barnsley and S. Demko (1985) [1] we introduce the following special class of dynamical systems, that has been named by Barnsley and Demko iterated function systems (IFS). Let (X, d) be a compact metric space. We start by considering a set of N single-valued continuous $w_i : X \to X$.

Then we construct the dynamical system IFS, i.e. the quadruple $(\mathcal{K}(X), h, \mathbf{Z}^+, W)$ on the set $\mathcal{K}(X)$ of all non-empty closed subsets of X, by defining the mapping $W : \mathcal{K}(X) \times \mathbf{Z}^+ \to \mathcal{K}(X)$ for each $S \in \mathcal{K}(X)$ as follows,

$$W(S) = \bigcup_1^N \hat{w}_i(S) := W(S, 1),$$

where $\hat{w}_i(S) := \{w_i(x) : x \in S\}$, and

$$W(S, 0) := S,$$

$$W(S, n+1) := W(W(S), n) \qquad n = 1, 2, \dots.$$

Because of the continuity of the maps w_i, $i = 1, 2, \dots, N$, we can extend the domain of W from $\mathcal{K}(X) \times \mathbf{Z}^+$ to $\mathcal{K}(X) \times \mathbf{Z}$ and the semi-dynamical system turns into a dynamical system.

At this point we assume the maps w_i, $i = 1, 2, \dots, N$, contractive on (X, d) that is for some real number s, with

$$0 \le s < 1$$

and all $x, y \in X$,

$$d(w_i(x), w_i(y)) \le s \, d(x, y), \qquad i = 1, 2, \dots, N.$$

With M.F. Barnsley and others (1986) [2], the corresponding IFS is called hyperbolic and the number s is called the contractivity factor of such dynamical system.

The following theorem states a peculiar common property of the hyperbolic IFS's.

Theorem 1. $W : \mathcal{K}(X) \to \mathcal{K}(X)$ is a contraction mapping with

$$h(W(A), W(B)) \leq s\, h(A, B) \quad \text{for all} \quad A, B \in \mathcal{K}(X).$$

Proof: By definition (4.1) of the Hausdorff distance

$$
h(W(A), W(B)) \;=\; \max \left\{ \underset{x \in \bigcup_1^N \hat{w}_i(A)}{\text{supremum}} \; \underset{y \in \bigcup_1^N \hat{w}_j(B)}{\text{infimum}} \; d(x, y), \right.
$$

$$
\left. \underset{x \in \bigcup_1^N \hat{w}_i(B)}{\text{supremum}} \; \underset{y \in \bigcup_1^N \hat{w}_j(A)}{\text{infimum}} \; d(x, y) \right\}
$$

$$
=\; \max \left\{ \underset{x \in A, 0 \leq i \leq N}{\text{supremum}} \; \underset{y \in B, 0 \leq j \leq N}{\text{infimum}} \; d(w_i(x), w_j(y)), \right.
$$

$$
\left. \underset{x \in B, 0 \leq i \leq N}{\text{supremum}} \; \underset{y \in A, 0 \leq j \leq N}{\text{infimum}} \; d(w_i(x), w_j(y)) \right\}
$$

$$
\leq\; \max \left\{ \underset{x \in A, 0 \leq i \leq N}{\text{supremum}} \; \underset{y \in B}{\text{infimum}} \; d(w_i(x), w_i(y)), \right.
$$

$$
\left. \underset{x \in B, 0 \leq i \leq N}{\text{supremum}} \; \underset{y \in A}{\text{infimum}} \; d(w_i(x), w_i(y)) \right\}
$$

$$
\leq\; \max \left\{ \underset{x \in A}{\text{supremum}} \; \underset{y \in B}{\text{infimum}} \; s \cdot d(x, y), \; \underset{x \in B}{\text{supremum}} \; \underset{y \in A}{\text{infimum}} \; s \cdot d(x, y) \right\}
$$

$$
=\; s \cdot h(A, B).
$$

By the contraction principle on the compact metric space $(\mathcal{K}(X), h)$. There exists in $\mathcal{K}(X)$ a unique set A (the attractor) defined by

$$A = \lim_{n \to \infty} W(S, n) \quad \text{for all} \quad S \in \mathcal{K}(X).$$

Moreover,

$$A = \lim_{n \to \infty} W(W(A), n - 1) = W(A), \tag{5.1}$$

A is indeed the (unique) fixed point of the contractive mapping $W : \mathcal{K}(X) \to \mathcal{K}(X)$. Set $A_i = w_i(A)$, $i = 1, 2, \ldots, N$, then (5.1) can be written as

$$A = \bigcup_1^N A_i \tag{5.2}$$

each of the A_i's is a (distorted) reduced copy of the attractor A, equation (5.2) says that the attractor has the property (self-tiling) of being the union of its copies (tiles) by the maps w_i, $i = 1, 2, \ldots, N$. Note that if the copies of X, i.e. X_i, $i = 1, 2, \ldots, N$, are two by two disjoint, so are the copies of any other set $S \in \mathcal{K}(X)$ and in particular those of the attractor. In this case the attractor is partitioned by its copies A_i, $i = 1, 2, \ldots, N$.

A **direct** problem in connection with this particular kind of dynamical systems could be: given the hyperbolic IFS, i.e. given an explicit representation for the contractive maps w_i, find the attractor, hoping that it turns out to look like the image of a familiar object.

We report here few classical examples [11], for each of them the mappings w_i, $i = 1, 2, \ldots, N$, are all contractive linear mappings. For additional examples we refer to M.F. Barnsley's book (1988) [3].

Ex. 1: $X = [0, 1] \subset \mathbb{R}$, $N = 2$, $w_1(x) = \frac{1}{3}x$, $w_2(x) = \frac{1}{3}x + \frac{2}{3}$,
A = ternary Cantor set.

'Ex. 2: $X = [0, 1] \times [0, 1] \subset \mathbb{R}^2$, $N = 2$

$$w_i(x) = A_i \begin{pmatrix} x_1 \\ x_2 \end{pmatrix} + b_i, \quad i = 1, 2, \ldots, N,$$

$$A_1 = \begin{pmatrix} \frac{1}{2} & \frac{1}{2} \\ -\frac{1}{2} & \frac{1}{2} \end{pmatrix} \quad b_1 = \begin{pmatrix} 0 \\ 0 \end{pmatrix}$$

$$A_2 = \begin{pmatrix} -\frac{1}{2} & \frac{1}{2} \\ -\frac{1}{2} & -\frac{1}{2} \end{pmatrix} \quad b_2 = \begin{pmatrix} 1 \\ 0 \end{pmatrix},$$

A = dragon.

Ex. 2: $X = [0, 1] \times [0, 1] \subset \mathbb{R}^2$, $N = 3$,

$$A_1 = \begin{pmatrix} \frac{1}{2} & 0 \\ 0 & \frac{1}{2} \end{pmatrix} \quad b_1 = \begin{pmatrix} 0 \\ 0 \end{pmatrix}$$

$$A_2 = \begin{pmatrix} \frac{1}{2} & 0 \\ 0 & \frac{1}{2} \end{pmatrix} \quad b_2 = \begin{pmatrix} \frac{1}{4} \\ \frac{1}{4}\sqrt{3} \end{pmatrix}$$

$$A_3 = \begin{pmatrix} \frac{1}{2} & 0 \\ 0 & \frac{1}{2} \end{pmatrix} \quad b_3 = \begin{pmatrix} \frac{1}{2} \\ 0 \end{pmatrix},$$

A = Sierpinski gasket.

Ex. 3: $X = [0, 1] \times [0, 1] \subset \mathbb{R}^2$, $N = 4$,

$$A = \begin{pmatrix} 0.856 & 0.041 \\ -0.0205 & 0.858 \end{pmatrix} \quad b = \begin{pmatrix} 0.07 \\ 0.147 \end{pmatrix}$$

$$A = \begin{pmatrix} 0.244 & -0.385 \\ 0.176 & 0.224 \end{pmatrix} \quad b = \begin{pmatrix} 0.393 \\ 0.102 \end{pmatrix}$$

$$A = \begin{pmatrix} -0.144 & 0.39 \\ 0.181 & 0.259 \end{pmatrix} \quad b = \begin{pmatrix} 0.527 \\ -0.014 \end{pmatrix}$$

$$A = \begin{pmatrix} 0 & 0 \\ 0.031 & 0.216 \end{pmatrix} \quad b = \begin{pmatrix} 0.486 \\ 0.05 \end{pmatrix},$$

A = Barnsley fern.

Ex. 4: $X = [0,1] \times [0,1] \subset \mathbb{R}^2, \quad N = 4,$

$$A = \begin{pmatrix} 0.8 & 0 \\ 0 & 0.8 \end{pmatrix} \qquad b = \begin{pmatrix} 0.1 \\ 0.04 \end{pmatrix}$$

$$A = \begin{pmatrix} 0.5 & 0 \\ 0 & 0.5 \end{pmatrix} \qquad b = \begin{pmatrix} 0.25 \\ 0.4 \end{pmatrix}$$

$$A = \begin{pmatrix} 0.355 & -0.366 \\ 0.355 & 0.355 \end{pmatrix} \qquad b = \begin{pmatrix} 0.266 \\ 0.078 \end{pmatrix}$$

$$A = \begin{pmatrix} 0.355 & 0.355 \\ -0.355 & 0.355 \end{pmatrix} \qquad b = \begin{pmatrix} 0.378 \\ 0.434 \end{pmatrix},$$

A = maple leaf.

Ex. 5: $X = [0,1] \times [0,1] \subset \mathbb{R}^2, \quad N = 8,$

$$A_1 = \begin{pmatrix} \frac{1}{3} & 0 \\ 0 & \frac{1}{3} \end{pmatrix} \qquad b = \begin{pmatrix} 0 \\ 0 \end{pmatrix}$$

$$A_2 = \begin{pmatrix} 0 & -\frac{1}{3} \\ \frac{1}{3} & 0 \end{pmatrix} \qquad b = \begin{pmatrix} \frac{1}{3} \\ \frac{1}{3} \end{pmatrix}$$

$$A_3 = \begin{pmatrix} \frac{1}{3} & 0 \\ 0 & \frac{1}{3} \end{pmatrix} \qquad b = \begin{pmatrix} 0 \\ \frac{2}{3} \end{pmatrix}$$

$$A_4 = \begin{pmatrix} 0 & \frac{1}{3} \\ -\frac{1}{3} & 0 \end{pmatrix} \qquad b = \begin{pmatrix} \frac{1}{3} \\ 1 \end{pmatrix}$$

$$A_5 = \begin{pmatrix} 0 & \frac{1}{3} \\ -\frac{1}{3} & 0 \end{pmatrix} \qquad b = \begin{pmatrix} \frac{1}{3} \\ \frac{1}{3} \end{pmatrix}$$

$$A_6 = \begin{pmatrix} \frac{1}{3} & 0 \\ 0 & \frac{1}{3} \end{pmatrix} \qquad b = \begin{pmatrix} \frac{2}{3} \\ 0 \end{pmatrix}$$

$$A_7 = \begin{pmatrix} 0 & -\frac{1}{3} \\ \frac{1}{3} & 0 \end{pmatrix} \qquad b = \begin{pmatrix} 1 \\ \frac{1}{3} \end{pmatrix}$$

$$A_8 = \begin{pmatrix} \frac{1}{3} & 0 \\ 0 & \frac{1}{3} \end{pmatrix} \qquad b = \begin{pmatrix} \frac{2}{3} \\ \frac{2}{3} \end{pmatrix},$$

A = Sierpinski carpet.

But the practical problem of representing an image by an IFS is exactly the opposite of what we have considered as the direct problem. The image of an object to be memorized in a computer is given in the form of an array (the raster) of $m \times n$ points (the pixels), white points for the background, black for the object (the foreground).

For data compression memorizing the image pixel by pixel is not the most efficient way of storing data. Since storing a computer program requires much less memory space, if one can replace the point by point storing by an algorithm that when needed produces the black pixels of the image, the data compression could be remarkable, and the resolution unlimited. This is what one tries to achieve by solving an inverse problem for IFS. Namely one considers the image (the target) as the attractor of an IFS, to be found within a certain class of mappings w_i. For the sake of the best data compression the mappings that are used are linear transformations.

From the point of view of the **existence** of an approximate solution to this inverse problem in the class of all hyperbolic IFS, a general positive answer is provided by the following theorems 2 and 3 (M.F. Barnsley and others (1986) [2]).

Theorem 2: (Geometric collage theorem) Let $(\mathcal{K}(X), h, W)$ be a hyperbolic IFS with contractivity factor s. Let S be a subset of $\mathcal{K}(X)$ such that

$$h(S, W(S)) \leq \epsilon, \tag{5.3}$$

for some $\epsilon > 0$. Then

$$h(S, A) < \epsilon/(1 - s),$$

where A is the attractor of the IFS.

Proof:

$$h(S, W(S, n)) \leq h(S, W(S, 1)) + h(W(S, 1), W(S, 2) + \cdots +$$

$$h(W(S, n-1), W(S, n))$$

$$\leq h(S, W(S, 1)) + s\, h(S < W(S, 1) + \cdots + s^{n-1} h(S, W(S, 1))$$

$$= \left[\sum_0^{n-1} s^i\right] h(S, W(S)) = \frac{1 - s^n}{1 - s} h(S, W(S)),$$

on taking the limit as $n \to \infty$, we obtain

$$h(S, A) \leq \frac{1}{1 - s} h(S, W(S)) \leq \epsilon/(1 - s).$$

This shows that given $\eta > 0$ and the target S in $\mathcal{K}(X)$, if we can find a hyperbolic IFS such that

$$h(S, W(S)) < \eta\,(1 - s)$$

then the corresponding attractor A differs from S (in Hausdorff distance) by less than η,

$$h(S, A) < \eta.$$

Theorem 3: The class $\mathcal{A} = \{A \in \mathcal{K}(X) : A$ is the attractor of some hyperbolic IFS$\}$ is dense in $(\mathcal{K}(X), h)$.

In other words any set S in $\mathcal{K}(X)$ can be approximated by the attractor of a hyperbolic IFS, within any desired order of approximation. We omit the proof of theorem 3 since we will report the proof of a more general result in the following paragraph.

6. IMAGE REPRESENTATION AND FUZZY SEMI-DYNAMICAL SYSTEMS

An image by a picture in black and white is a finite set P of points in an $m \times n$ array. Usually these points (the pixels) are not just black (foreground) or white (background) but to each pixel p_{ij} a **grey level** or brightness value is attached represented by a number h in the interval $[0,1]$ ($0 =$ white, $1 =$ black). The function $h : P \to [0,1]$ that defines the grey level distribution over the pixels, can have two opposite functions. It can provide information on the extension of the third dimension (or depth) of the objects in the picture or it can represent the noise, as in the blurred picture of an object out of focus. Let us stick to the first function. When we transfer the image on the computer we would like to keep the information which is provided by the image function h. That is, we expect the computer to keep in storage the function h or better to be able to reproduce the function h when needed. Now we can see through the image function h that the natural mathematical representation of a picture is by means of a fuzzy set $u : X \to [0,1]$. In representing an image and its "informative" grey level distribution no probabilistic meaning is attached to the value $u(x)$ at each point $x \in X$. On the contrary, when the grey level distribution is the "noise" then $u(x)$ can be interpreted as the probability of x to be a pixel of the object (foreground).

The pixels in the image P can be classified according to their grey level by considering for each α, with $0 < \alpha \leq 1$, the set $[u]^{\alpha}$ of all the pixels whose grey level is not less than α. Then the closure $[u]^0$ of the set $\{x \in X : u(x) > 0\}$ is the region in the picture that is covered by the image of the object, while the remaining area $X \backslash [u]^0$ is definitely the background. One recognizes immediately in the sets $[u]^{\alpha}$ the level sets of the fuzzy set u. Note that there is a one-to-one correspondence between an image function (fuzzy set) and the set $\{[u]^{\alpha} : 0 \leq \alpha \leq 1\}$ of all the α-level sets.

Following C. Cabrelli, B. Forte and others (1992) [9], we now introduce a particular class of fuzzy semi-dynamical systems that has been successfully used to represent an image with its grey level distribution. The method has also been used to represent a colour picture; the fuzzy set u is then replaced with a fuzzy 4-vector $u = (u_1, u_2, u_3, u_4)$, where $u_i(x)$, $i = 1, 2, 3$, is the percentage of the basic colour i at x, $u_4(x)$ is the measure of its brightness.

To define our particular class of fuzzy semi-dynamical system on the given complete metric space $(\mathcal{F} * (X), d_{\infty})$ we have to choose a single valued mapping $\hat{W} : \mathcal{F} * (X) \to \mathcal{F} * (X)$.

With the given compact metric space (X, d), we consider, as for the IFS, N contractive, single-valued, continuous mappings $w_i : X \to X$.

Parallel to the introduction of the set-valued mappings $\hat{w}_i$, for each $u \in \mathcal{F} * (X)$ and each subset S of X we define

$$\tilde{u}(S) := \text{supremum}\,\{u(y) : y \in S\} \quad \text{if} \quad S \neq \phi,$$

$$\tilde{u}(\phi) := 0,$$

that implies in particular $\tilde{u}(\{x\}) = u(x)$ for each $x \in X$. Then for each w_i, $i = 1, 2, \ldots, N$, we define

$$\tilde{u}_i(x) := \tilde{u}(\hat{w}_i^{-1}(\{x\})),$$

where clearly $\hat{w}_i^{-1}(\{x\}) = \phi$ if $x \notin w(X)$. Each $\tilde{u}_i$, $i = 1, 2, \ldots, N$, is a fuzzy set, indeed it maps $X \to [0, 1]$. Moreover, with $u \in \mathcal{F}*(X)$, there exists $x_0 \in X$ such that $u\hat{w}_i x_0) = 1$. For each i, choose y_{i0} so that $w_i(x_0) = y_{i0}$, then $\tilde{u}(\hat{w}_i^{-1}(\{y_{i0}\})) = 1$. Hence the fuzzy sets $\tilde{u}_i$, $i = 1, 2, \ldots, N$, are normal.

Moreover, each $\tilde{u}_i$ is upper semi-continuous, or equivalently, for each $i = 1, 2, \ldots, N$, and each $0 \le \alpha \le 1$, the level sets

$$C_i^\alpha := \{x \in X : \tilde{u}_i \ge \alpha\}$$

are closed. For $\alpha = 0$, since $C_i^0 = X$, all the C_i^0 are trivially closed. Now assume $\alpha > 0$, because of the compactness of X and the upper semi-continuity of u_i, if $C_i^\alpha \ne \phi$, than $x \in C_i^\alpha$ implies that there exists $y \in X$ such that $w_i(y) = x$ and then there exists for each x_n a y_n in X such that $w_i(y_n) = x_n$ and $u(y_n) \ge \alpha$. By the compactness of X there exists a subsequence $\{y_{n_k}\}$ of $\{y_n\}$ which is convergent to a limit $\bar{y}$ in X.

By the continuity of $w_i : X \to X$,

$$\lim_{k \to \infty} w_i(y_{n_k}) = w_i(\bar{y}),$$

which implies

$$\bar{x} = \lim_{k \to \infty} x_{n_k} = w_i(\bar{y}).$$

But by the upper semi-continuity of u

$$u(\bar{y}) \ge \lim_{k \to \infty} \sup u(y_{n_k}) \ge \alpha,$$

which shows that $\bar{x}$ is indeed in C_i^α. Thus for each $i = 1, 2, \ldots, N$, and all $0 \le \alpha \le 1$, C_i^α is closed, which completes the proof that $u \in \mathcal{F}*(X)$ implies $\tilde{u}_i \in \mathcal{F}*(X)$, $i = 1, 2, \ldots, N$.

Finally, to each fuzzy set $\tilde{u}_i$ a grey level map $\phi_i : [0, 1] \to [0, 1]$ is associated. These maps have the role to modify the value of $\tilde{u}_i(x)$, that is, the grey levels of each pixel, and to complete the definition of the fuzzy semi-dynamical system we set

$$\tilde{W}(u)(x) := \text{supremum} \{\phi_1(\tilde{u}_1(x))), \phi_2(\tilde{u}_2(x)), \ldots, \phi_N(\tilde{u}_N(x))\}$$
$$= \tilde{W}_s u \tag{6.1}$$

To guarantee that this mapping maps indeed $\mathcal{F}*(X)$ into $\mathcal{F}*(X)$, we require that the following conditions be met,

1) each ϕ_i is non-decreasing on $[0, 1]$,

2) each ϕ_i is right continuous on $[0, 1)$,

3) $\phi_i(0) = 0$ for all $i = 1, 2, \ldots, N$, and

4) for at least one $j \in \{1, 2, \ldots, N\} : \phi_j(1) = 1$.

For the sake of simplicity we define

$$q_i(x) = \phi_i(\tilde{u}_i(x)), \qquad i = 1, 2, \ldots, N.$$

Then we have,

Lemma 1: The functions $q_i : X \to [0, 1]$ exhibit the following properties

(a) each q_i is upper semi-continuous,

(b) for each $\alpha \in [0,1]$ and $i = 1, 2, \ldots, N$, $\left[q_i\right]^\alpha = \hat{w}_i\left([\phi_i \circ u]^\alpha\right)$,

(c) for each $\alpha \in [0,1]$,

$$[\tilde{W}_s u]^\alpha = \bigcup_1^N [q_i]^\alpha = \bigcup_1^N \hat{w}_i\left([\phi_i \circ u]^\alpha\right),$$

(remember that $\hat{w}_i(S) := \{w_i(x) : x \in S\}$).

Proof: (a) is a direct consequence of conditions 1 and 2 on ϕ_i and the proved upper semi-continuity of $\tilde{u}_i$.

To prove property (b) we separate the two cases $\alpha = 0$ and $\alpha > 0$. With $\alpha = 0$, because of conditions 1 and 3 on ϕ_i, it is easy to recognize that

$$\{x \in X : \phi_i(\tilde{u}_i(x)) > 0\} = \hat{w}_i(\{x \in X : (\phi_i \circ u)(x) > 0\}),$$

hence
$$[q_i]^0 = \text{closure}\,\{x \in X : \phi_i(\tilde{u}_i(x)) > 0\}$$

$$= \overline{\hat{w}_i(\{x \in X : (\phi_i \circ u)(x) > 0\})}$$

and by the continuity of w_i,

$$\overline{\hat{w}_i(\{x \in X : (\phi_i \circ u)(x) > 0\})}$$

$$= \hat{w}_i(\overline{\{x \in X : (\phi_i \circ u)(X) > 0\}})$$

$$= \hat{w}_i([\phi_i \circ u]^0).$$

For $\alpha > 0$ the property is a straightforward consequence of conditions 1 and 2 on ϕ_i

To prove (c) note that if $x \in [\tilde{W}_s u]^\alpha$ than for at least one i $\{1, 2, \ldots, N\}$, $x \in [q_i]^\alpha$ otherwise $\tilde{W}(u)(X) < \alpha$ by the definition (6.1); conversely if, for a $j \in \{1, 2, \ldots, N\}$, $x \in [q_j]^\alpha$ then $\phi_j(\tilde{u}_j(x)) \geq \alpha$ and again by definition (6.1), $\tilde{W}(u)(x) \geq \alpha$, thus: $x \in [\tilde{W}_s u]^\alpha$.

The fuzzy semi-dynamical system defined by the quadruple $(\mathcal{F} * (X), \ d_\infty, \ \mathbf{Z}^+, \ \tilde{W}_s)$ completed with the set $\phi = \{\phi_1, \phi_2, \ldots \phi_N\}$ of functions $\phi_i : [0,1] \to [0,1]$, has been called in Cabrelli, B. Forte and others (1992) [9] the iterated fuzzy set system (IFZS).

The main feature of such a system is contained in theorem 4 (C. Cabrelli, B. Forte and others (1992) [9]). However, to make the proof of this theorem easier we use the following Lemma 2, whose proof we shall omit (see again C. Cabrelli, B. Forte and others (1992) [9], Appendix A).

Lemma 2: If $\phi : [0,1] \to [0,1]$ is non-decreasing right continuous, and $u \in \mathcal{F}(X)$ is upper semi-continuous then there exists a decreasing sequence $\{\beta_n\}$ in $(0,1)$ such that $[\phi \circ u]^0 = \lim_{n \to \infty} [u]^{\beta_n}$ in Hausdorff metric.

Theorem 4: The mapping $\tilde{W}_s$ on $(\mathcal{F} * (X), d_\infty)$ as defined in (6.1) is a contractive mapping $\mathcal{F} * (X) \to \mathcal{F} * (X)$, that is for each $u \in \mathcal{F} * (X)$,

$$\tilde{W}_s u \in \mathcal{F} * (X) \quad \text{and}$$

with $0 \leq s < 1$,

$$d_\infty(\tilde{W}_s u, \tilde{W}_s v) \leq s\, d_\infty(u, v)$$

for all $u, v, \in \mathcal{F} * (X)$.

Proof: Since for each $\alpha \in [0, 1]$, $[q_i]^\alpha$ are closed sets, their union $[\tilde{W}_s u]^\alpha$ (see Lemma 1) is a closed set, which proves that $\tilde{W}_s u$ is upper semi-continuous. By property (4) of the functions ϕ_i, there exists $j \in \{1, 2, \ldots, N\}$ such that $\phi_j(1) = 1$, but all the $\tilde{u}_i$ are normal fuzzy sets, thus there exists $x_0 \in X$ such that $\tilde{u}_j(x_0) = 1$, hence $q_j = \phi_j(\tilde{u}_j(x_0)) = 1$ which implies $(\tilde{W}_s u)(x_0) = 1$, by definition (6.1). So we have proved that for each $u \in \mathcal{F} * (X)$, the fuzzy set $\tilde{W}_s u$ is also upper semi-continuous and normal; therefore $\tilde{W}_s u \in \mathcal{F} * (X)$.

To prove that the mapping $\tilde{W}_s$ is continuous on $(\mathcal{F} * (X), d_\infty)$ we introduce the following notation. For each $\alpha \in (0, 1]$ we denote with P_α the set of all i in $\{1, 2, \ldots, N\}$ such that $\phi_i(1) \geq \alpha$, and then with P_0 the set of all i in $\{1, 2, \ldots, N\}$ such that $\phi_i(1) > 0$. Clearly, since the fuzzy sets $u \in \mathcal{F} * (X)$ are all normal, if $i \in P_\alpha$ then there exists $u \in \mathcal{F} * (X)$ such $[\phi_i \circ u]^\alpha \neq \phi$, more than that: for all $u \in \mathcal{F} * (X): [\phi_i \circ u]^\alpha \neq \phi$ and vice versa.

Then we start with

$$d_\infty(\tilde{W}_s u, \tilde{W}_s v) = \operatorname*{supremum}_{0 \leq \alpha \leq 1} \{h([\tilde{W}_s u]^\alpha, [\tilde{W}_s v]^\alpha)\} \tag{6.2}$$

where (see 4.1))

$$h([\tilde{W}_s u]^\alpha, [\tilde{W}_s v]^\alpha) = \max\{D([\tilde{W}_s u]^\alpha, [\tilde{W}_s v]^\alpha), D([\tilde{W}_s v]^\alpha, [\tilde{W}_s u]^\alpha)\}. \tag{6.3}$$

We can write

$$\begin{aligned}
D([\tilde{W}_s u]^\alpha, [\tilde{W}_s v]^\alpha) &= D\left(\cup_{i \in P_\alpha} \hat{w}_i([\phi_i \circ u]^\alpha), \cup_{j \in P_\alpha} \hat{w}_j([\phi_j \circ v]^\alpha)\right) \\
&= \max_{i \in P_\alpha} D(\hat{w}_i([\phi_i \circ u]^\alpha), \cup_{j \in P_\alpha} \hat{w}_j([\phi_j \circ v]^\alpha)) \\
&\leq \max_{i \in P_\alpha} D(\hat{w}_i([\phi_i \circ u]^\alpha), w_i([\phi_i \circ v]^\alpha)) \\
&\leq s \max_{i \in P_\alpha} D([\phi_i \circ u]^\alpha, [\phi_i \circ v]^\alpha)
\end{aligned}$$

and similarly for $D([\tilde{W}_s v], [\tilde{W}_s u])$ in (6.3)

$$D([\tilde{W}_s v]^\alpha, [\tilde{W}_s u]^\alpha) \leq s \max_{i \in P_\alpha} D([\phi_i \circ v]^\alpha, [\phi_i \circ u]^\alpha),$$

hence,

$$h([\tilde{W}_s u]^\alpha, [\tilde{W}_s v]^\alpha) \leq s \max_{i \in P_\alpha} h([\phi_i \circ u]^\alpha, [\phi_i \circ v]^\alpha).$$

For each $i = 1, 2, \ldots, N$ and $\alpha \in (0, \phi_i(1)]$, there exists a $\beta(i)$, namely $\beta(i) =$ infimum $\{t : \phi_i(t) \geq \alpha\}$, such that

$$[\phi_i \circ u]^\alpha = [u]^{\beta(i)}.$$

On the other hand for each i in P_0, $\phi_i(1) > 0$ and for $u \in \mathcal{F} * (X)$, $[\phi_i \circ u]^0 \neq \Phi$, and by Lemma 2 there exists a decreasing sequence $\{\beta_n(i)\}$ in $(0, 1)$ such that

$$[\phi_i \circ u]^0 = \lim_{n \to \infty} [u]^{\beta_n},$$

in Hausdorff distance. Then

if $0 < \alpha \leq \phi_i(1)$, i.e. $i \in P_\alpha$, $h([\phi_i \circ u]^\alpha, [\phi_i \circ v]^\alpha = h([u]^{\beta(i)}, [v]^{\beta(i)}) \leq d_\infty(u, v)$,

if $\alpha = 0$ and $i \in P_0$, $h([\phi_i \circ u]^0, [\phi_i \circ v]^0) = \lim_{n \to \infty} h([u]^{\beta_n(i)}, [v]^{\beta_n(i)}) \leq d_\infty(u, v)$,

thus

$$h([\tilde{W}_s u]^\alpha, [\tilde{W}_s v]^\alpha) < s\, d_\infty(u, v) \quad \text{for all} \quad \alpha \in [0, 1],$$

hence

$$d_\infty(\tilde{W}_s u, \tilde{W}_s v) \leq s\, d_\infty(u, v).$$

The mapping $\tilde{W}_s$ on $(\mathcal{F} * (X), d_\infty)$ is contractive. By the Contraction Mapping Principle on the complete metric space $(\mathcal{F} * (X), d_\infty)$ we have

Proposition 1. For each IFZS there exists a unique fuzzy set $u* \in \mathcal{F} * (X)$ (the attractor) such that

$$\tilde{W}_s u* = u * .$$

In other words in $\mathcal{F} * (X)$ there exists unique a solution $u*$ to the functional equation in the unknown $u(x)$

$$u(x) = \text{supremum} \{\phi_1(\tilde{u}(w_1^{-1}(x))), \phi_2(\tilde{u}(w_2^{-1}(x))), \ldots, \phi_N(\tilde{u}(w_N^{-1}(x))), \quad (6.4)$$

for all $x \in X$.

The name attractor for the fuzzy set $u*$ is justified by the following.

Proposition 2. For each $v \in \mathcal{F} * (X)$

$$d_\infty(\tilde{W}_s(v, n), u*) \to 0 \quad \text{as} \quad n \to \infty.$$

Moreover by Lemma 1 we have

Proposition 3. (Generalized self tiling) $u*$ is the attractor of the IFZS $(\mathcal{F}^*((X), d_\infty, \mathbf{Z}^+, \tilde{W}_s, \phi)$ if and only if for all $\alpha \in [0, 1]$

$$[u*]^\alpha = \bigcup_1^N \hat{w}_i([\phi_i \circ u*]^\alpha).$$

The following proposition illustrates how the sequence of fuzzy sets $\tilde{W}_s(v, n)$ converges to the attractor u.

But first we remind the following definition.

Definition For $u, v \in \mathcal{F} * (X)$ we shall say that $u \leq v$ if and only if $u(x) \leq v(x)$ for all $x \in X$.

Then, calling the **base IFS** of the given IFZS $(\mathcal{F}*(X), d_\infty, \mathbf{Z}^+, \tilde{W}_s, \phi\}$ the quadruple $(\mathcal{K}(X), h, \mathbf{Z}^+, \hat{W})$ where $\hat{W}$ is the contraction map as defined in paragraph 4 by the contraction maps $\hat{w}_i$, $i = 1, 2, \ldots, N$, we have

Proposition 4. Let $u*$ be the attractor of a given IFZS and A the attractor of its base IFS, for each $v \in \mathcal{F} * (X)$ and $B \in \mathcal{K}(X)$

1a - $\tilde{W}_s v \leq v$ implies $u* \leq v$,

1b - $W(B) \subseteq B$ implies $A \subseteq B$,

2a - $v \leq \tilde{W}_s v$ implies $v \leq u*$,

2b - $B \subseteq W(B)$ implies $B \subseteq A$.

We leave the proof of this proposition to the reader, who can, however, find it in C. Cabrelli, B. Forte and others (1992) [9].

The connection between the fuzzy set attractor $u*$ and the attractor will be pointed out by the following theorem.

Theorem 5: If $u*$ is the fuzzy set attractor of the IFZS $(\mathcal{F} * (X), d_\infty, \mathbf{Z}^+, \tilde{W}_s, \phi)$ and A is the attractor of its base IFS then

$$\text{support } u* = [u*]^0 \subseteq A. \qquad (6.4)$$

Proof: Let χ_A be the characteristic function of A. The attractor A being an invariant set ($\hat{W}A = A$), $x \notin A$ implies $w_i^{-1}(x) \subset X \backslash A$, for all $i = 1, 2, \ldots, N$. That is

$$\tilde{W}_s \chi_A \leq \chi_A,$$

and by 1a in Proposition 4,

$$u* = \lim_{n \to \infty} \tilde{W}_s(\chi_A, n) \leq \chi_A,$$

which yields

$$[u*]^0 \subseteq A.$$

This means that in the representation of an image by IFZS, the foreground is not larger than the foreground in the corresponding representation by the base IFS. Equality in (6.4) holds for sure in the following two cases

1 - when for all $i = 1, 2, \ldots, N : \phi_i(1) = 1$,

2 - when for all $i = 1, 2, \ldots, N : \phi_i(u) > 0$ for all $u > 0$.

In fact

$$[u*]^0 = \bigcup_1^N \hat{w}_i([\phi_i \circ u*]^0) = \bigcup_1^N \hat{w}_i(A) = \hat{W}(A) = A.$$

However, the real life problem does not require to find the fuzzy set attractor of a given IFZS, but it requires just the opposite, that is given an image, represented via the image function by a fuzzy set v, find an IFZS such that v coincides or almost coincides with its attractor $u*$. Notice that to solve such an inverse problem we have now more flexibility than when we were dealing with the IFS. Indeed, we have now more variables to be determined namely the contraction mapping $\tilde{W}_s : \mathcal{F}*(X) \to \mathcal{F}*(X)$ and the set $\phi = \{\phi_1, \phi_2, \ldots, \phi_N\}$, of functions $\phi_i : [0,1] \to [0,1]$. We are also going to provide more information about the object in the picture by preserving the grey level distribution. The existence of an approximate solution to the (inverse) problem is guaranteed by the following Theorems 6 and 7.

Theorem 6: (IFZS Geometric collage theorem, C. Cabrelli, B. Forte and others (1992) [9]). Given an IFZS with contractivity factor s, if for some $\epsilon > 0$, there exists $v \in \mathcal{F}*(X)$ such that

$$d_\infty(v, \tilde{W}_s v) \leq \epsilon, \tag{6.5}$$

then

$$d_\infty(v, u*) < \epsilon(1-s),$$

where $u*$ is the attractor of the IFZS.

Proof: Except for the necessary changes in the symbols the proof is identical to that of theorem 2. In fact, we start with

$$d_\infty(v, \tilde{W}_s(v,n)) \leq d_\infty(v, \tilde{W}_s(v,1)) + d_\infty(\tilde{W}_s(v,1), \tilde{W}_s(v,2)) + \cdots +$$

$$d_\infty(\tilde{W}_s(v, n-1), \tilde{W}_s(v,n))$$

$$\leq d_\infty(\tilde{W}_s(v,1)) + s \cdot d_\infty(v, \tilde{W}_s(v,1)) + \cdots s^{n-1} \cdot d_\infty(v, \tilde{W}_s(v,1))$$

$$= \left[\sum_0^{n-1} s^i\right] d(v, W(v,1)) = \frac{1-s^n}{1-s} d_\infty(v, \tilde{W}_s v),$$

and on taking the limit as $n \to \infty$ we get

$$d_\infty(v, u*) \leq \frac{1}{1-s} d_\infty(v, Wv) \leq \epsilon/(1-s).$$

This proves that if we are looking for an IFZS such that its attractor $u*$ differs from a given fuzzy set $v \in \mathcal{F}*(X)$, the image, by less than a given $\eta > 0$, in the d_∞-metric, then it is sufficient to find an IFZS such that

$$d_\infty(v, \tilde{W}_s v) \leq \eta \cdot (1-s).$$

Theorem 7: (C. Cabrelli and U. Molter (1992) [10]). If (X, d) is a compact metric space and $(\mathcal{F} * (X), d_\infty)$ is the complete metric space of all normal and upper semi-continuous fuzzy sets on (X, d), as defined in paragraph 4, then the subclass of fuzzy sets

$$\mathcal{U} := \{u* \in \mathcal{F} * (X) : u* \text{ is the attractor of some IFZS on } (\mathcal{F} * (X), d_\infty)\}$$

is dense in $(\mathcal{F} * (X), d_\infty)$.

Proof: It will be achieved through four steps:

Step 1 - Given $v \in \mathcal{F} * (X)$ and $\epsilon > 0$, we construct the N mappings $w_i : X \to X$, as follows. By the compactness of $[v]^0$ there exists a finite covering of that set made of N open balls $B_i = B\left(x_i, \frac{\epsilon}{4}\right)$, $i = 1, 2, \ldots, N$, where

$$B(x, r) := \{y \in X : d(x, y) < r\}.$$

Thus

$$[v]^0 \subset \bigcup_1^N B_i. \tag{6.6}$$

Then we choose N contractive maps $w_i : X \to X$ with contraction factor $s \leq \frac{1}{2}$, $i = 1, 2, \ldots, N$, and $\hat{w}_i(X) \subset B_i$, $i = 1, 2, \ldots, N$.

Step 2 - Set $\alpha_0 = 0$ and $\alpha_i = \text{supremum } \{v(x) : x \in \overline{B}_i\}$, where $\overline{B}_i$ is the closure of B_i. Then for each $\alpha \in [0, 1]$, we have

$$[v]^\alpha \subset \bigcup_{\{i : \alpha \leq \alpha_i\}} B_i. \tag{6.7}$$

We choose the set $\phi = \{\phi_1, \phi_2, \ldots, \phi_N\}$ of functions $\phi_i : [0, 1] \to [0, 1]$ so that

1) each ϕ_i is non-decreasing right continuous on $[0, 1]$,

2) $\phi_i(1) = \alpha_i$ (this together with (1) implies $\phi_i(x) \leq \alpha_i$), $i = 1, 2, \ldots, N$.

As an example, we could choose $\phi_i = \alpha_i \chi_{[\alpha_i, 1]}$, where $\chi_{[\alpha_i, 1]}$ is the characteristic function of the closed interval $[\alpha_i, 1]$. Thus we have

$$[\phi_i \circ u]^\alpha \neq \phi \quad \text{if} \quad \alpha \leq \alpha_i$$

$$[\phi_i \circ v]^\alpha = \phi \quad \text{if} \quad \alpha > \alpha_i.$$

Then

$$[\tilde{W}_s v]^\alpha = \bigcup_1^N w_i([\phi_i \circ v]^\alpha) = \bigcup_{\{i : \alpha \leq \alpha_i\}} w_i([\phi_i \circ v]^\alpha). \tag{6.8}$$

Step 3 - We denote with $D_d(S)$, $d > 0$, $S \in \mathcal{K}(X)$, the "inflated" version of S

$$D_d(S) := \{x \in X : d(x, S) < d\}.$$

Then one can easily verify that

$$[v]^\alpha \subset \bigcup_{\{i:\alpha \le \alpha_i\}} B_i \subset D_{\frac{\epsilon}{2}}([v]^\alpha) \tag{6.9}$$

But $w_i(S) \subset B_i$ and because of the particular choice of the approximations ($\epsilon/4$ and then $\epsilon/2$),

$$w_i(S) \subset B_i \subset D_{\frac{\epsilon}{2}}(w_i(S)), \tag{6.10}$$

for all $S \in \mathcal{K}(X)$ and $i = 1, 2, \ldots, N$.

By recourse to (6.8), (6.10) yields

$$[\tilde{W}_s v]^\alpha \subset \bigcup_{\{i:\alpha \le \alpha_i\}} B_i \tag{6.11}$$

and since for each $i = 1, 2, \ldots, N$, and $\alpha \le \alpha_i$

$$[\tilde{W}_s v]^\alpha \cap B_i \ne \phi, \qquad [v]^\alpha \cap B_i \ne \phi,$$

(6.7) and (6.10) yield

$$[v]^\alpha \subset \bigcup_{\{i:\alpha \le \alpha_i\}} B_i \subset D_{\frac{\epsilon}{2}}([\tilde{W}_s v]^\alpha) \tag{6.12}$$

$$[\tilde{W}_s v] \subset \bigcup_{\{i:\alpha \le \alpha_i\}} B_i \subset D_{\frac{\epsilon}{2}}([v]^\alpha).$$

But for each $B, C \in \mathcal{K}(X)$,

$$B \subset D_d(C) \qquad \text{and} \qquad C \subset D_d(B) \qquad \text{imply}$$

$$h(B, C) \le d,$$

then

$$h([v]^\alpha, [\tilde{W}_s v]^\alpha) \le \epsilon/2,$$

hence

$$d_\infty(v, \tilde{W}_s v) \le \epsilon/2. \tag{6.13}$$

Step 4 - By recourse to (the Collage) theorem 6, with $s = \max\{s_i, i = 1, 2, \ldots, N\} \le 1/2$ and (6.13) we have

$$d_\infty(v, u*) \le \epsilon/[2(1 - s)] \le \epsilon/[2(1/2)] = \epsilon,$$

where $u*$ is the attractor of the IFZS that has been constructed through steps 1 and 2. Since v is an arbitrary fuzzy set in $\mathcal{F} * (X)$, the existence of such $u*$ proves that the set of the attractors of all IFZS is dense in $(\mathcal{F} * (X), d_\infty)$.

7. IFS COMPARED TO IFZS

The following proposition shows that the choice of the mapping $\tilde{W}_s$ with definition (6.1) was not arbitrary, if we want the IFZS to be consistent with its base IFS, which is defined through the mappings $\hat{w}_i : (X) \rightarrow X\mathcal{K}(X)$ by the map (tiling)

$$\hat{W}(S) := \bigcup_1^N \hat{w}_i(S), \quad \text{for each} \quad S \in \mathcal{K}(X). \tag{7.1}$$

Proposition 5 A mapping $\tilde{W} : \mathcal{F}*(X) \rightarrow \mathcal{F}*(X)$ exhibits the tiling property for the level set $[u]^\alpha$ of each $u \in \mathcal{F}*(X)$, for all $\alpha in[0,1]$, that is

$$[\tilde{W}u]^\alpha = \bigcup_1^N \hat{w}_i([\phi_i ou]^\alpha), \tag{7.2}$$

if and only if $\tilde{W} = \tilde{W}_s$, that is if and only if (see (6.1))

$$(\tilde{W}u)(x) = \text{supremum} \{\phi_1(\tilde{u}_1(x)), \phi_2(\tilde{u}_2(x)), \dots, \phi_N(\tilde{u}_N(x))\}, \tag{7.3}$$

with $\tilde{u}_1(x) := \tilde{u}(w_i^{-1}(x))$, $i = 1, 2, \dots, N$.

Proof: By Lemma 1 in paragraph 6 we know that (7.3) implies (7.2). Because of the one to one correspondence between the set of level sets $\{[u]^\alpha : \alpha \in [0,1]\}$ and fuzzy sets $u \in \mathcal{F}*(X)$, the unique fuzzy set $\tilde{W}u$ defined by (7.2) (right to left) must coincide with the one given by (7.3).

The similarity between the IFS and the IFZS approach is even more stringent as one can realize by following step by step the definitions of the two algorithms. If we proceed on strictly parallel paths, we have:

1 - With (X, d) a compact metric space and $\mathcal{K}(X)$ the set of all non-empty closed sets in X, we introduce the Hausdorff metric h on $\mathcal{K}(X)$, which becomes a complete metric space, then we consider a set of N continuous contractive maps $w_i : X \rightarrow X$, $i = 1, 2, \dots, N$,

IFS	IFZS

IFS

2 - we define the set to set maps $\hat{w}_i$: $\mathcal{K}(X) \to \mathcal{K}(X)$ as follows

$$\hat{w}_i(S) := \{w_i(x) : x \in S\}.$$

3 - all $\hat{w}_i$ are contractive maps on $(\mathcal{K}(X), h)$.

4 - we define the operator $\hat{W}$: $\mathcal{K}(X) \to \mathcal{K}(X)$ as follows

$$\hat{W}(S) := \bigcup_1^N \hat{w}_i(S) \qquad (7.4)$$

for each $S \in \mathcal{K}(X)$.

5 - this operator $\hat{W}$ is contractive on $(\mathcal{K}(X), h)$.

IFZS

2 - we define, using the extension principle for fuzzy sets

$$\tilde{\tilde{w}}_i(u) = v \in \mathcal{F} * (X),$$

where $v(x) := \sup\{u(y) : y = w_i^{-1}(x)\}$.

3 - for each i, $\tilde{\tilde{w}}_i$ is a contractive map on $(\mathcal{F} * (X), d_\infty)$.

4 - we define the operator $\tilde{\tilde{w}} : \mathcal{F} * (X) \to \mathcal{F} * (X)$ as follows

$$(\tilde{\tilde{w}}u)(x) := \sup_{1 \le i \le N} \{\tilde{\tilde{w}}_i(u(x))\} \qquad (7.5)$$

for each $u \in \mathcal{F} * (X)$.

5 - this operator is contractive on $(\mathcal{F} * (X), d_\infty)$.

Since $\sup\{\chi_C(x), \chi_D(X)\} = \chi_{C \cup D}$, the restriction of the operator (7.5) to the subclass of $\mathcal{F} * (X)$ of the characteristic functions of all sets S in $\mathcal{K}(X)$ coincides with (7.4), with $\chi_{\hat{w}(s)} := \tilde{\tilde{w}} \chi_S$. Note that (7.5) is the mapping $\tilde{W}_S : \mathcal{F} * (X) \to \mathcal{F} * (X)$ associated to an IFZS $(\mathcal{F} * (X), d_\infty, \mathbf{Z}^+, \tilde{W}, \phi)$ when all ϕ_i in $\phi = \{\phi_1, \phi_2, \ldots, \phi_N\}$ are identity maps: $\phi_i(t) = t$ for all $t \in [0,1]$. In this case, as in the more general case where $\phi_i(1) = 1$, $i = 1, 2, \ldots, N$, the attractor of the IFZS if the characteristic function of the attractor of its base IFS. Thus if we define the IFZS on $\mathcal{F} * (X)$ by closely translating into the language of fuzzy sets the IFS approach on $\mathcal{K}(X)$ we do not make any more progress in image representation than what we already made by the IFS approach.

What provides a useful generalization with added flexibility is the introduction of the set of function ϕ into the definition of the mapping $\tilde{W}$. This even with respect to the probabilistic approach where a set $P = \{p_1, p_2, \ldots, p_N\}$ of probabilities (replacing ϕ) $p_i > 0$ are attached to the maps w_i (see M.F. Barnsley and others (1985) [1]). Besides nothing prevents us from adding flexibility by replacing the set of functions $\phi_i : [0,1] \to [0,1]$ with a set $\phi = \{\phi_1, \phi_2, \ldots, \phi_N\}$ where $\phi_i : X \times [0,1] \to [0,1]$.

On the other hand one could solve the problem of representing the image function via an algorithm using straight IFS's.

To explain this point let us assume that our target is a continuous grey level distribution v over $X = [0,1]$ (a more realistic case would be $X = [0,1] \times [0,1]$). We can consider as our new target the closed subset of $[0,1] \times [0,1]$

$$S := \{(x,y) \in \mathbb{R}^2 : 0 \le x \le 1, \quad 0 \le y \le v(x)\}$$

and approximate it by the attractor A of an IFS. (See Fig. (7.1))

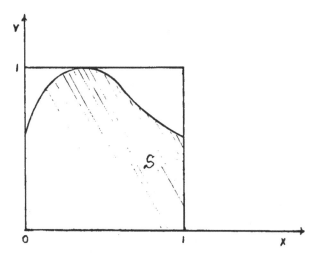

Figure 7.1

The price that we have to pay for a simpler algorithm (IFS vs. IFZS) is in computer time for an anneeded information (the pixels in the interior of S).

Another promising application of the IFZS algorithm is in the graphic display of a function, that has an explicit analytic representation: see for instance the function

$$u = \sin \pi x \qquad 0 \le x \le 1,$$

or through a set of a sufficient number of data with a sufficient accuracy.

Indeed, every function which is bounded and upper semi-continuous on a compact subset X of $\mathbb{R}$ can be reduced (by an affine transformation) to a function $u : [0, 1] \to [0, 1]$ that is to a fuzzy set u in $\mathcal{F} * ([0, 1])$.

8. CHAOTIC DYNAMICAL SYSTEMS AND IMAGE REPRESENTATION

The generation of an image by a chaos game provides a useful example of an inverse problem for chaotic dynamical systems (see R. Devaney (1986) [6]). The corresponding algorithm to generate the image is a recurrent iterated function system (RIFS) (see M.F. Barnsley and others (1988) [4], (1989) [5]). The setting is the same as that used in defining an IFS. On a compact metric space (X, d), N continuous contractive maps $w_i : X \to X$, $i = 1, 2, \ldots, N$, are given; in addition to each map w_i a probability $p_i > 0$, $\sum_i^N p_i = 1$, is associated. In the probability distribution $P = \{p_1, p_2, \ldots, p_N\}$, p_i represents the (stationary) probability that at any stage in a sequence of trials the map w_i is picked up. This model can also be viewed as the model of an autonomous semi-dynamical system whose solution operator $w(x, t)$, $x \in X$, $t \in \mathbf{Z}^+$, is random.

Its randomness being due to the fact that the image of (x, n) is given by any w_i through $w((x,n),n) = w_i((x,n))$ and therefore is not uniquely determined; p_i is the probability that the particle starting from (x,n) is moved in the unit time interval by the map w_i to $w_i((x,n))$.

Let $\mathcal{B}(X)$ be the σ-algebra of Borel subsets of X and $\mathcal{P}(B)$ the set of all probability measures on $\mathcal{B}$, i.e. $\nu \in \mathcal{P} : \mathcal{B} \to [0,1]$, $\nu(X) = 1$. A metric (Hutchinson metric) d_H has been defined on $\mathcal{P}$, namely

$$d_H(\mu, \nu) := \sup_{f \in L_1} \left[\int_X f d\mu - \int_X f d\nu \right] \quad \text{for all} \quad \mu, \nu \in \mathcal{P}.$$

where $L_1 := \{ f : X \to \mathbb{R} :| \text{ f(x)} - \text{f(y) } | < \text{d(x,y)} \}$. The metric space $(\mathcal{P}, d_H)$ is complete (J. Hutchinson (1981) [15]).

Associated with the contractive continuous maps $w_i : X \to X$, $i = 1, 2, \ldots, N$, and the set of probabilities $P = \{ p_1, p_2, \ldots, p_N \}$ consider the map (Markow operator) $M : \mathcal{P} \to \mathcal{P}$, which is defined as follows

$$M(\nu) := \sum_1^N p_i \nu \circ \hat{w}_i^{-1}, \tag{8.1}$$

where as usual $\hat{w}_i(B) = \{ w_i(x) : x \in B \}$.

Parallel to Theorems 1 and 4 we have

Proposition 6: The map M is contractive on $(\mathcal{P}, d_H)$, i.e.

$$d_H(M\mu, M\nu) \le s \, d_H(\mu, \nu) \quad \text{for all} \quad \mu, \nu \in \mathcal{P}.$$

As a direct consequence we have

Proposition 7 There exists a unique probability measure $\mu*$ in $\mathcal{P}$ such that

$$M\mu * (S) = \mu * (S) \quad \text{for all} \quad S \in \mathcal{B}.$$

Starting with any probability measure $\nu \in \mathcal{P}$, this invariant $\mu*$ can be obtained as a limit according to

$$\mu * (S) = \lim_{n \to \infty} M(\nu(S), n) \quad \text{for all} \quad S \in \mathcal{B}, \tag{8.2}$$

where $M(\nu(S), 0) := \nu(S)$, $M(\nu(S), 1) := M(\nu(S))$ and $M(\nu(S), n+1) := M(\nu(S), n)$ for all $n = 1, 2, \ldots$.

The metric space $(\mathcal{P}, d_H)$ together with the Markow operator M, i.e. the probabilities $P = \{ p_1, p_2, \ldots, p_N \}$ and, $\hat{w} = \{ \hat{w}_1, \ldots, \hat{w}_N \}$ thus the quadruple $(\mathcal{P}, d_H, \mathbb{Z}^+, M)$, defines a (probabilistic) **chaotic semi-dynamical system**. A meaningful relation between a chaotic semi-dynamical system and its base IFS is the content of the following

Proposition 8: If A is the attractor of the base IFS then support $(\mu*) = A$, that is

$$\mu * (S) = \mu * (A \cap S) \quad \text{for all} \quad S \in \mathcal{B}(X). \tag{8.3}$$

The proof of this proposition is based on the contactiveness of the Markow operator M on $(\mathcal{P}, d_H)$, (see J. Hutchinson (1981)).

A collage theorem will conclude this set of properties of a semi-dynamical system.

Theorem 8: (Collage theorem for measures) If for a given $\epsilon > 0$, there exists $\nu \in \mathcal{P}(\mathcal{B})$ such that

$$d_H(\nu, M\nu) \leq \epsilon,$$

then

$$d_H(\nu, \mu*) \leq \epsilon/(1-s),$$

where s is the contractivity factor of the maps $w_i : X \rightarrow X$, in (8.1), $i = 1, 2, \ldots, N$.

The proof of this theorem is identical to that for theorem 6 and it shall be omitted.

Given the image of an object (a piece of hardware, a tool, a piece of furniture, etc.) in black and white one can consider its digitized version as the result of the discretization of a continuous grey level distribution. Such distribution may be represented by a (probability) measure ν on the σ-algebra of Borel subsets of a compact set X (a rectangle when dealing with a photo). In this model the grey level of a pixel will be the measure of the set which is represented by that pixel. Then the object in the picture and the grey level distribution on it can be identified with the attractor of an appropriate chaotic semi-dynamical system. Thus the direct problem associated with this model reads as follows, given the image by the semi-dynamical system $(\mathcal{P}, d_H, \mathbf{Z}^+, M)$ find the attractor $\mu*$ (the object). This direct problem can be solved by recourse to an iteration algorithm. One starts with a unit point mass measure at x_0,

$$\nu_0(S) = \delta_{x_0}(S) = \begin{cases} 0 & \text{if } x_0 \notin S \\ 1 & \text{if } x_0 \in S \end{cases},$$

where x_0 is preferably a point of the attractor A of the base IFS. Then by (8.1) we have

$$\nu_1 = p_1 \delta_{w_1(x_0)} + p_2 \delta_{w_2(x_0)} + \cdots + p_N \delta_{w_N(x_0)}$$

$$\nu_n = \sum_{i_1 i_2, \ldots, i_n} p_{i_1} p_{i_2} \cdots p_{i_n} \, \delta_{w_{i_1} \circ w_{i_2} \circ \cdots \circ w_{i_n}(x_0)}$$

where $\{i_1, i_2, i_3, \ldots, i_n\}$ is any sequence of n numbers from the set $\{1, 2, \ldots, N\}$. If $\mu*$ is the probability measure attractor of the semi-dynamical system, then we know by (8.2) that $d_H(\mu*, \nu_n) \rightarrow 0$ as $n \rightarrow \infty$. In other words we have in ν_n an approximate solution to the direct problem.

Another (fast) approach to solve the direct problem is the random iterated function system or chaos game. To explain it we need some preliminary definitions and remarks.

Let $\sum_N$ be the set of all half-infinite strings of symbols in the alphabet that consists of the N integers $\{1, 2, \ldots, N\}$,

$$\sigma \in \sum_n \Rightarrow \sigma = \{\sigma_1, \sigma_2, \ldots, \sigma_n, \ldots\}$$

$$\sigma_i \in \{1, 2, \ldots, N\}, \qquad i = 1, 2, \ldots, n, \ldots .$$

Define for each couple of sequences $\sigma, \tau \in \Sigma_N$, a distance d_{Σ_N} as follows,

$$d_{\Sigma_N}(\sigma, \tau) := \sum_{k=1}^{\infty} \frac{|\sigma_k - \tau_k|}{(N+1)^k}.$$

The couple $\left(\Sigma_N, d_{\Sigma_N}\right)$ is a compact metric space (see R. Devaney (1986) [6]). For each $x_0 \in X$, define the random n-steps path $g(\sigma, n, x_0)$ through the compositions

$$g(\sigma, 0, x_0) := x_0, \ g(\sigma, 1, x_0) := w_{\sigma_1}(x_0), \ g(\sigma, 2, x_0) := w_{\sigma_2} \circ w_{\sigma_1}(x_0),$$

$$g(\sigma, n, x_0) := w_{\sigma_n} \circ w_{\sigma_{n-1}} \circ w_{\sigma_{n-2}} \circ \cdots \circ w_{\sigma_1}(x_0),$$

with $\sigma = \{\sigma_1, \sigma_2, \ldots, \sigma_n, \ldots\}$,
and let

$$g(\sigma, x) := \lim_{n \to \infty} (\sigma, n, x).$$

the existence of such limit for all $x \in X$ is a direct consequence of the following basic inequality

Proposition 9: If $D = \operatorname{diam}(X) = \sup\{d(x', x'') : x', x'' \in X\}$ then

$$d(g(\sigma, m, x'), g(\sigma, n, x'')) \le s^{\min\{m,n\}} D, \tag{8.4}$$

for all $m, n \in Z, \ x', x'' \in X$.

Proof: Assume, without any loss of generality, $m \le n$. Since

$$g(\sigma, n, x) = w_{\sigma_n} \circ \cdots \circ w_{\sigma_{m+1}}(w_{\sigma_m} \circ \cdots \circ w_{\sigma_1}(x)),$$

we have with $\ y'_m = w_{\sigma_m} \circ w_{\sigma_{m-1}} \circ \cdots \circ w_{\sigma_1}(x'), \ \ y''_m = w_{\sigma_m} \circ w_{\sigma_{m-1}} \circ \cdots \circ w_{\sigma_1}(x'')$

$$d(g(\sigma, m, x'), g(\sigma, n, x'')) \le \sup_{x', x''} d(y'_m, w_n \circ w_{n-1} \circ \cdots \circ w_{m+1}(y''_m))$$

$$s^m \sup_{x', x''} d(x', w_n \circ w_{n-1} \circ \cdots \circ w_{m+1}(x'')) \le s^m D$$

Proposition 10:

a) $g(\sigma, x) := \lim_{n \to \infty} g(\sigma, n, x)$ exists and the convergence of $g(\sigma, n, x)$ to $g(\sigma, x)$ on (X, d) is uniform,

b) $g(\sigma, x)$ belongs to the attractor A of the base IFS and $g(\sigma, x)$ does not depend on x,

c) the map $g : \Sigma_N \to A$, is continuous on $\left(\Sigma_N, d_{\Sigma_N}\right)$ and onto and finally

d) if $w_i(A) \cap w_j(A) = \phi$, for all $i, j \in \{1, 2, \ldots, N\}, \ i \ne j$, then $g : \Sigma_N \to A$ is one-to-one.

With the product probability

$$p_{\sigma_1} p_{\sigma_2} \cdots p_{\sigma_n}$$

assigned to the (cylinder) set of all strings σ in $\sum_N$ with the same first n symbols $\sigma_1, \sigma_2, \ldots, \sigma_n$, a unique probability measure P is defined on the σ-algebra of parts of $\sum_N$ that is generated by the said (cylinder) sets.

The random iteration algorithm or chaos game (an effective algorithm for solving the direct problem, given a random semi-dynamical system) is essentially based on the following theorems.

Theorem 9 Let

$$x_{n+1} = w_{\sigma_n}(x_n) \qquad n = 0, 1, 2, \ldots,$$

where the sequence $\sigma = \{\sigma_1, \sigma_2, \ldots\}$ is chosen randomly in $(\sum_N, P)$ i.e. the indices σ_n are chosen randomly and independently form $\{1, 2, \ldots, N\}$ with probabilities prob $(\sigma_n = i) = p_i$, then with probability 1 the set $\{x_0, x_1, \ldots, x_n, \ldots\}$ (the orbit) is dense on A (the attractor).

This means that with probability 1, for every $\epsilon > 0$ and $y \in A$, there exists $\bar{n} \in \mathbf{Z}^+$, such that $d(x_{\bar{n}}, y) < \epsilon$.

Moreover, if $\mu*$ is the invariant measure (i.e. $\mu* = M \mu*$, the attractor) in $\mathcal{P}(\mathcal{B})$ of the given semi-dynamical system then

Theorem 10: For each given $x_0 \in X$, and almost all sequences σ in $\left(\sum_N, d_{\sum_N} \right)$

$$\lim_{n \to \infty} \frac{1}{n+1} \sum_0^n \chi_B(x_h) = \mu^*(B),$$

for all $B \in \mathcal{B}(\chi)$, and with χ_B the characteristic function of B.

For the proof of these theorems we refer the reader to J.H. Elton (1987) [13].

The quantity

$$\frac{1}{n+1} \sum_0^n \chi_B(x_h) \tag{8.5}$$

is the ratio of the number of times x is in B along the path $\{x_0, x_1, \ldots, x_n\}$ to the total number of states in that path or equivalently the frequency of visitation of B relative to that path.

The chaos game is the natural algorithm that follows directly from Theorems 9 and 10. We shall describe the main steps of this algorithm.

Step 1. We chose a random number generator for sequences (uniformly) distributed on $[0, 1]$,

Step 2. We choose as first point of a path (the seed) one of the fixed points $w_j(x_0) = x_0$ of one of the N maps w_i, $i = 1, 2, \ldots$. Certainly this point x_0 is in A as well as all the points of any trajectory that starts at x_0.

Step 3. Having partitioned the interval $[0, 1]$ into N subintervals by the points $\{t_0 = 0 < t_1 < \cdots < t_n = 1\}$ with $t_{i+1} - t_i = p_i$, $i = 0, 1, \ldots, n - 1$. If the n-th number produced by the random number generator is in the interval $[t_i, t_{i+1})$ we choose w_i to evaluate $x_{n+1} = w_{\sigma_n}(x_n)$, $n = 0, 1, 2, \ldots$.

Step 4. Each pixel represents indeed a set B of points of the image, before its digitization, one can evaluate by (8.5) the frequency of visitation of B along the given path and choose accordingly the grey level for that pixel.

The **inverse problem** (image representation) can be stated as follows:

Given the measure $\mu*$ (the attractor) on $(\mathcal{P}, d_H)$, find a chaotic semi-dynamical contractive system $(\mathcal{P}, d_H, \mathbf{Z}^+, M)$ that has such measure $\mu*$ as invariant measure (attractor).

Theorem 8 (collage theorem for meaures) suggests a way to solve this problem; for if ν is our target in $(\mathcal{P}, d_H)$ and $\epsilon > 0$ is a given number (an upper bound for the error) it is sufficient to find $(w_1, w_2, \ldots, w_N)$ and $(p_1, p_2, \ldots, p_N)$ so that

$$d_H(\nu, M\nu) \leq \epsilon (1 - s)$$

in order to have

$$d_H(\nu, \mu*) \leq \epsilon,$$

hence to have in the chaotic semi-dynamical system $(\mathcal{P}, d_H, \mathbf{Z}^+, M)$ with attractor $\mu*$ an approximate solution to the problem.

As it concerns the existence of such an approximate solution we have the following Theorem (C. Cabrelli, private communication) in the particular case $X = [0, 1]$.

Theorem With $X = [0, 1]$ and $\mathcal{M} * (\mathcal{B})$ the set of all measures which are attractors for some chaotic semi-dynamical system on $\mathcal{P}(\mathcal{B})$, $\mathcal{M} * (\mathcal{B})$ i dense in $(\mathcal{P}(\mathcal{B}), d_H)$, i.e. for each $\nu \in \mathcal{P}(\mathcal{B})$ and each $\epsilon > 0$ there exists $\mu* \in \mathcal{M} * (\mathcal{B})$ such that $d_H(\nu, \mu*) \leq \epsilon$.

9. CONCLUDING REMARKS

We hope that we have given enough motivations that can stimulate more research in the field of inverse problems. We have seen very real life problems waiting for a satisfactory solution that cannot be obtained but by solving inverse problems. We have seen how formal solutions to typical problems in classical one particle dynamics, do not provide a satisfactory answer to the most common problem: knowing some solutions find all solutions in a certain specific class of solutions. Finally, we have seen that for very practical purposes more research is needed in the resuscitated field of general dynamical systems. In particular, dynamical systems on function spaces seem to provide a very powerful tool to model real life situations of interest for industries and the applied sciences in general. With the application of fuzzy dynamical systems to represent images we gave here just a modest example.

REFERENCES

1. Barnsley M.F., Demko S., "Iterated Function Systems and the Global Construction of Fractals", *Proc. Roy. Soc. London Ser. A* **399**, 243-275 (1985).

2. Barnsley M.F., Ervin V., Hardin D., and Lancaster J., "Solution of an Inverse Problem for Fractals and Other Sets", *Proc. Nat. Acad. Sci. USA*, vol. 83, 1975-1977 (1986).

3. Barnsley M.F., **Fractals Everywhere**, Academic Press Inc., San Diego, CA (1988).

4. Barnsley M.F., Sloan A.D., "A Better Way to Compress Images", *BYTE Magazine*, Jan. Issue, 215-223 (1988).

5. Barnsley M.F., Elton J. and Hardin D.P., "Recurrent Iterated Function Systems", *Constr. Approx.* **B5**, 5-31 (1989).

6. **Devaney R., An Introduction to Chaotic Dynamical Systems, Addison Wesley (1986).**

7. Birkhoff G.D., "Dynamical Systems", *AMS Colloquium Publications*, vol. 9 (1927), revised edition (1966).

8. Galiullin A.S., **Inverse Problems of Dynamics**, Mir Publishers (1984).

9. Cabrelli C.A, Forte B., Molter U.M. and Vrscay E.R., "Iterated Fuzzy Set Systems: A New Approach to the Inverse Problem for Fractals and Other Sets", to appear in *J. Math. Anal. Appl.* (1992).

10. Cabrelli C.A., Molter U.M., "Density of Fuzzy Attractors: A Step Towards the Solution of the Inverse Problem for Fractals and Other Sets", manuscript.

11. Dubuc S., Elqortobi A., "Approximations of Fractal Sets", *J. Comput. and Appl. Math.* **29**, 79-89 (1990).

12. Gladwell G.M.L., "Lectures on Inverse Problems", n. 1, *SASIAM Reports*, Bari Tecnopolis (1989).

13. Elton J.H., "An Ergodic Theorem for Iterated Maps", *Ergod. Th. & Dynam. Sys.* **7**, 481-488 (1987).

14. Harvey G.F., "Mathematical Simulation of Tight Coil Annealing", *J. Aust. Inst. Metals* **22**, 1, 28-37 (1977).

15. Hutchinson J., "Fractals and Self-similarity", *Indiana Univ. Math. J.* **30**, 713-747 (1981).

16. Kloeden P.E., "Fuzzy Dynamical Systems", *Fuzzy Sets and Systems* **7**, 275-296 (1982).

17 Lavrentiev M.M., "Some Improperly Posed Problems of Mathematical Physics", *Springer Tracts in Natural Philosophy*, vol. 11, C. Truesdell ed., Springer-Verlag (1967).

18. Meshcherskii I.V., "Works on the Mechanics of Bodies with Varying Mass", *Gostekhizdat*, Moskow, Leningrad (1949), (in Russian).

19. Mundie D., "A Mathematical Model of the Batch Annealing Process", Master's Thesis, University of Waterloo, Dept. of Applied Mathematics (1981).

20. Stikker U.O., "Numerical Simulation of the Coil Annealing Process, Mathematical Models in Metallurgical Process Development", *ISI pub.*, **123**, 104-114 (1970).

21. Suslov G.K., "On a Force Function Admitting Given Integrals", Kiev, (1890), (in Russian).

22. Zadeh L.A., "Fuzzy Sets", *Inform. Control* **8**, 338-353 (1965).

23. Diamond P., Kloeden P.E., "Metric Spaces of Fuzzy Sets", *Fuzzy Sets and Systems*, vol. 35, 241-249 (1990).

Case Studies in Industrial Mathematics

Stavros Busenberg

Department of Mathematics, Harvey Mudd College, Claremont, CA 91711 USA

1 University–Industry Collaborations in Applied Mathematics

Mathematics is a vital component of the technological base that supports modern economies and societies. This is not to say that this is all that mathematics is or should be. However, the field has an important social function that adds to the forces that pull the discipline away from a self-centered narrow view of itself as a pure art form with a limited elite audience. Those mathematicians who accept this social responsibility find it imperative to seek interactions and collaborations with industry in order to promote the effectiveness of the mathematical sciences in helping technological enterprises. But such a quest soon leads to the realization that there are very few natural settings in which direct interactions between mathematicians and industry can occur. The obstacles to these interactions are often due to the fact that the vast majority of research mathematicians are employed by university departments whose structure is designed to reward individual research work. Such a reward system often leads to a suspicious view of contacts with commercial and industrial entities. The isolation of the academic mathematician has also lead to cliché misconceptions about the value of applied and industrial mathematics on the part of the academics, which have been reciprocated by the opinion of some industrial mathematical scientists that the entire pure mathematical enterprise is worthless. These radical inclinations have found their celebrated proponents on both extremes. Even though those who subscribe to these prejudices are not always individuals of limited talent, these viewpoints invariably indicate a marked narrowness of intellectual perception. While recognizing the existence of these views, we shall be concerned here only with the bridges that have been constructed to facilitate university-industry collaborations and with some samples of problems that have been addressed in such a setting. In doing this we will be following the tradition of mathematics in the spirit of most of its outstanding creators. The vast majority of truly great mathematicians such as Archimedes, Newton, Euler, and Gauss, and from this century Hilbert, von Neumann, Kolmogorov, and Wiener, to name only a few, have vigorously pursued both mathematics and its varied applications.

Over the past quarter century the lack of natural venues for University-Industry interactions in mathematics has been recognized by a number of individuals who have worked to address this problem. Each particular setting has required different ways of providing a structure to support these interactions. In this section I will give brief descriptions of some of these programs in order to provide examples of what has been done, of the ingredients that they have in common, and of how local conditions tend to lead to different types of structures. The Mathematics Clinic program in Claremont is the setting with which I have been intimately connected since my colleagues and I started it at Harvey Mudd College in 1973. Because of my familiarity with this program, I will describe it first and use it as a basis for comparison with other major university-industry programs that have been flourishing for some years.

The central idea of the Mathematics Clinic program is that the practice of industrial mathematics is best learned through an apprenticeship experience which simulates the type of activity the mathematician is likely to encounter in modern industrial settings. Work-study programs, in which undergraduate students spend part of one or more of their undergraduate academic years working for some group in industry, have had a long tradition. However, they have the dual disadvantages of interrupting the academic work of the students while placing the students in industrial settings where they are often delegated rather low level tasks because of the lack of educational programs and the time pressures that occur in industry.

In designing the Mathematics Clinic, we addressed both of the drawbacks of work-study programs. First, we brought the industrial problems and their sponsors to our campus through a structured academic program that allowed the students to continue with their other studies while being involved in industrial work. Second, we charged the sponsoring industrial firms a realistic price for the work on their problem, thus ensuring that only serious problems would be presented to the Mathematics Clinic and that there would be sufficient stake in the outcome of the project for the industrial liaison to place time pressures and professional level expectations on the students involved in the program.

The Mathematics Clinic started at Harvey Mudd College with a single project that I recruited from the Bell and Howell company in 1973, and whose technical content will be described in a later section. A team of three students worked with me on this project, and a physicist at the Bell and Howell laboratories, Denis Rose, was the industrial liaison. This first project was quite successful and the following year the Clinic program was expanded at Harvey Mudd College and also adopted by the Claremont Graduate School as an integral part of its Master's degree program in applied mathematics. At this writing the Mathematics Clinic program is part of the applied mathematics curricula at all of the Claremont Colleges, and has an average of six to seven industrial projects per year, each involving three to five students, a faculty supervisor, and a liaison from industry. The fee per project paid by each company was $8,300 in 1973 and has risen to $31,500 at the present (1991).

For the students each Clinic project is a course in which they may enroll, but it is a course that is operated in a very untraditional manner (see [1-4] for detailed descriptions.) There are no lectures, no textbooks, no preset curriculum, no homework assignments, and no examinations. This is, of course, exactly the same as in the work faced by mathematical scientists in industry. Having settled the main negatives of the course, let us turn to what it does require. First, the students (three to five per project), the faculty supervisor, and the liaison are regarded as a team working on the same project. The efforts of this team are directed at the problem that is brought and funded by the industrial sponsor, and all of the mathematics that is learned and applied is directed to the understanding and solution of this specific problem. Second, as is invariably the case with industrial work, the team is expected to make periodic professional level oral and written reports. Two major written milestones are the midterm report at the end of the first semester in December, and the final report which is submitted to the sponsoring company at the end of the academic year in May. Three oral presentations are given by the students, one during each of the semesters and one at the end of the academic year. The audience at these presentations consists of industrial scientists, faculty and other students, and very high standards are expected.

The typical Clinic project comes to the team as a problem that is stated almost invariably in non-mathematical terms. For example, the first clinic project was stated as a problem encountered in rear projection screens which led to an undesirable scintillation that appeared to be a random diffraction pattern, but which could not be explained by either experimental observations or by the standard application of the theory of optics. The sponsor asked the Clinic team to develop a theoretical explanation of this phenomenon, which had substantial commercial implications, so that the experimentalists could focus on methods for dealing with it. The Clinic team had to start with the job of formulating an appropriate mathematical model for this problem. After a considerable amount of discussion with scientists and engineers at Bell and Howell, the team read suggested background material, considered a variety of possible approaches, and did some rough calculations on various simple models. Based on this background work, a model was formulated. The liaison and his colleagues were convinced that this model had the potential for adequately explaining the phenomenon. The team then concentrated on analyzing the model, using various simplifications dictated by either the physical setting or by mathematical necessity. In the process, the students needed to acquire a substantial number of mathematical tools and results that they had not encountered previously, but which were needed for the analysis of the model. The team also had to do a substantial amount of reading in the literature of optics and of some engineering articles concerned with the related problems of diffraction from random arrays of atmospheric particles and of laser speckles. In the end, the analysis and computer simulations of the Clinic team provided an explanation of the scintillation phenomenon and several suggestions on possible approaches for eliminating it. The client was sufficiently convinced of the value of this work that the team was requested not to disclose its findings for a period of a year!

The process that was described in the previous paragraph is to a large degree typical for the average Clinic project. The problem comes stated in non-mathematical terms. It is then modeled appropriately after an initial exploratory stage that involves literature searches and conversations with experts. The model is mathematically analyzed, often by the use of methods that have to be acquired or developed during the process of analysis. Almost invariably, numerical algorithms are developed and solved using computers in order to provide numerical tests of specific cases of interest to the sponsor. Finally, the work of the team is presented both orally and in formal written reports submitted to the client.

The students clearly learn mathematical and management skills that are not even hinted upon in traditional courses. They need to cultivate substantial communications skills in order to function well as a team as well as in order to provide acceptable oral and written reports to the client. They acquire experience in attacking an open-ended problem whose mathematical content and solution are not predetermined. They see first hand the expectations that are placed on industrial mathematical scientists, and experience the excitement of using their mathematical skills in providing useful information for a problem which is of value to industry. In my own personal experience as a faculty supervisor on clinic projects starting from the very inception of this program, I have concluded that, on the average, the students also learn much more mathematics than they do in traditional courses with comparable time requirements. This is largely so because the mathematics that they need to learn and to immediately apply does not come artificially compartmentalized by some textbook author or lecturer. Rather it is motivated by an immediate application which requires that the underlying concepts be understood thoroughly enough to be applied in a nontrivial manner.

The Clinic projects have encompassed a very broad spectrum of areas. Among these have been the modeling and analysis of physical problems, problems from statistics and optimization often arising from management and quality control programs, Ecological and resource management problems which have required a vast array of analytical and numerical methods, economic and sociological problems, and a variety of computer related problems. The following are the topics and the sponsors of the projects undertaken during the 1990-1991 academic year. The first four were done at Harvey Mudd College, the fifth at the Claremont Graduate School and the sixth at Claremont McKenna College.

1. *Automating the Analysis of Photogrametric Data*, sponsored by Hughes Simulation Systems.

2. *Prediction of Proton Densities in the Lower Van Allen Belt*, sponsored by the McDonnell Douglas Aircraft Company.

3. *Evaluation of Chaotic and Non-monotone Behavior in Combat Models*, sponsored by the RAND Coporation.

4. *Statistical Process Control of Wire-Semiconductor Bonding Processes*, sponsored by Teledyne Microelectronics.

5. *Heat Generation and Dissipation in Transistors*, sponsored by the Information Sciences Institute.

7. *Statistical Analysis of Software Data*, sponsored by the Jet Propulsion Laboratory.

We now turn briefly to another notable example of a long term university industry collaboration in applied mathematics, the Oxford Study Groups. These are annual week-long meetings of about a dozen industrial scientists and about fifty faculty and graduate students. The goals of these study groups are to help industrial research workers solve current industrial problems, to involve academic mathematicians in these important practical problems, and to provide a setting for introducing and training graduate students in industrial mathematics.

The structure of these meetings is designed to concentrate attention on particular problems. These problems are solicited and selected by the organizer (John Ockendon has played this role for many years) after personal consultations with industrial scientists and mathematicians. Typically the problems and models come from the physical sciences and require methods of continuum mechanics. On the first day of each meeting, the invited scientists from each of the industrial concerns make a short, about half hour, presentation of a current problem from their particular industry. Groups then form, consisting of faculty, graduate students and industrial scientist, which attack these problems. There is a lively exchange of ideas and much concentrated work on each of these specific problems. The last day is devoted to a presentation of the progress that was made during the week. The benefit for the industrial participants is, first of all, an immediate return of advice from experts who have concentrated their attention on their particular problem. On the longer term, there is the possibility that the week-long group effort will lead to future work by interested participants. On the academic side, these study groups often lead to thesis topics of current industrial interest and to interactions and collaborations with colleagues who would otherwise work in isolation. On the longer term, the organizers and others who are involved in these workshops for several years in a row can discern patterns and directions for new research programs and fields. One particularly noteworthy example of such a field that has been stimulated by the Oxford Study Groups is the topic of free boundary problems for partial differential equations, which recurred very often in industrial problems presented to these study groups. The paper [5] by Dewynne, Ockendon and Wilmott conveys some of the flavor of the types of problems that have been tackled by the Oxford Study Group. The Oxford example has served as a model for a number of other industrial mathematics activities such as the program at the Ransselear Polytechnic Institute.

There is also a strong tradition of University-Industry collaboration in mathematics at the University of Florence. The paper by Fasano and Primicerio [6] serves to illustrate the type of work that comes out of that program.

Of more recent vintage, and with a shorter but significant track record, is the program at the Institute for Mathematics and its Applications (IMA) of the University of Minnesota. The industrial mathematics activities at the IMA are described in a series of four books by Avner Friedman [7], and are largely concentrated in seminar type activities by visiting industrial scientists. A rather broad spectrum of problems of industrial interest are brought to the IMA in this way, and these are disseminated mainly through the above mentioned yearly volumes.

The rest of this paper will be devoted to short descriptions of three Clinic projects, one in each of the following sections. The format that the first and third of these descriptions will follow is one that parallels the typical process that the Clinic teams encounter and which involves the following steps:

1. A problem is presented in non-mathematical language, and the initial task is to understand what the client really wants and needs to know.

2. The problem is reformulated in the language of mathematics, that is, it is modeled. Here there are many crucial decisions that need to be made concerning what simplifications are valid, what aspects need to be kept at all cost, and what is an acceptable range of parameters for which the model needs to remain accurate. This stage often involves a number of sheer guesses which need to be tested by preliminary analyses and comparisons of predicted results with experimental observations.

3. The problem is analyzed mathematically. Several iterations of these first three steps may need to be taken before settling on the final model to be thoroughly studied. This part of the project is the one for which the traditional mathematics education comes closest to giving an adequate preparation.

4. Almost invariably, numerical computations of specific cases of particular interest to the sponsoring industry need to be made. In some projects this numerical aspect of the project may require considerable ingenuity and may easily rival the analytical aspects in the amount of effort and ingenuity that it requires.

5. The results, or rather parts of the results, are presented to the client in the correct language which almost invariably is not mathematical. Most often graphical presentations of these results are the most effective way of conveying them to the client.

In adopting this format for presenting two of the projects we will be giving up the temptation to abstract and beautify the pure mathematical aspects present in each of the problems that are described. We will also not attempt to present the current state of knowledge of each one of these problems, but rather describe the project as it evolved during the Clinic team's work on it. I hope that this type of description will convey some of the spirit of the work on these real world problems.

The format for the description of the second project differs, because here I wish to describe a process that has occurred frequently with these Clinic problems. That is, the stimulation of further work on an industrial problem, outside the specific format of the Clinic project that initiated it. The ties with the industrial application are, of course, very direct both because of the initial problem which inspires the work, and because of the personal contacts with industrial scientists who remain a source of information and encouragement during this second phase. Quite often this type of extension of the results of the Clinic stimulate future Clinic projects in the same area.

I have chosen three projects involving rather different types of mathematical models and methods to illustrate these processes. Each of these is a project in which I have been personally involved, and consequently, there is a limit to the range of the areas that they cover. In particular, I will not describe any project that is predominantly statistical in nature, even though about one fourth of the

Clinic work is of this type. This is so because of the limitations of my direct experience and not due to the lack of importance of these types of models and methods in industrial mathematics.

References

1. Borrelli, R., and S. Busenberg (1980): Undergraduate classroom experiences in applied mathematics, *The UMAP J.*, **1**, pp. 17-24.
2. Borrelli, R. and J. Spanier (1983): The Mathematics Clinic: a review of its first decade, *The UMAP J.*, **4**, pp. 1-28.
3. Busenberg, S. and W. Tam (1977): An academic program providing realistic training in software engineering, *Proc. of the ACM*, **9**, pp. 341-345.
4. Cumberbatch, E. and S. Busenberg (1986): Claremont's mathematics clinics complete twelfth year, *SIAM News*, January 1986.
5. Dewynne, J., J. Ockendon and P. Wilmott (1989): On a mathematical model for fiber tapering, *SIAM J. Appl. Math.*, **49**, pp. 983-990.
6. Fasano, A. and M. Primicerio (1991): Modelling the rheology of a coal-water slurry, *Proc. 4th Symp. on Math. in Industry, (ECMI 89)*, Hj. Wacker, Ed., Teubner-Kluwer, Amsterdam.
7. Friedman, A., (1988-1991): Mathematics in Industrial Problems, Parts 1-4, Springer-Verlag, New York.

2 Scintillations in Rear Projection Screens

2.1 Description of the Problem

In this section we will present a short description of the project that launched the Mathematics Clinic program in 1973. The students on the team where three fourth year mathematics majors at Harvey Mudd College, Joseph Costello, Stephen Quist and Frank Valdes, and I was the faculty supervisor. The sponsoring company was Bell and Howell. The problem was initially presented as follows:

Rear projection screens consisting of a glass surface that is coated with a layer of polymer material serve as the focussing surface for the images from a slide projected on the rear of the glass surface that is facing the viewer. When such screens are viewed for prolonged periods of time they lead to severe eye strain which, upon closer examination, is due to a multicolored pattern that the viewer sees on a plane that is close, but not identical, to the one where the projected image appears. This causes the eyes to involuntarily switch focus from the image plane to the pattern plane and back again, leading to eye fatigue. This undesirable pattern appears on close examination to be a collection of randomly distributed diffraction patterns that are not expected in this case since the light source is an incandescent lamp emitting incoherent radiation. An explanation of this phenomenon is needed which is sufficiently detailed to suggest avenues for eliminating it.

The experimental physicists at the Bell and Howell laboratories had examined this phenomenon in some detail and had determined that it was still present when viewed through polarizing lenses. They also had varied the light sources and the types of coatings on the glass screens without succeeding to remove the pattern. One of the puzzling things that they wanted explained was how such diffraction patterns could occur when the light source was an extended incandescent lamp filament and not a highly coherent source. Also, their intuitive arguments led to the conclusion that since the polymer screen coatings were randomly distributed on the glass surface, any diffraction patterns that could occur would likely be canceled out by their random superposition.

After considerable discussion with the physicists, it was decided to use the simplest model of this phenomenon that conserved the two crucial parameters consisting of an extended light source and of a polymer screen coating which produced a random variation in the index of diffraction. Since polarization did not affect the pattern, the scalar theory of light was deemed adequate. Consequently, the model that was adopted for detailed analysis consisted of a source of finite extent radiating light of a bounded spectral width which is lined up with a screen and a viewing plane. The screen was modeled as a random distribution of opaque disks (or else disks with index of diffraction $n > 1$) lying in a plane. The screen could alternately be modeled by assuming that it affected the phase and amplitude of the light impinging on it in some manner (either deterministic or random) that varied from point to point on the screen. The model of the rear projection screen that was used is depicted in Figure 2.1.

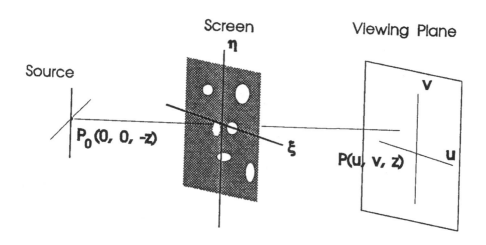

Fig. 2.1. Model of the rear projection screen

A model based on classical scalar theory of diffraction due to point sources can be used to show that the scintillation phenomenon is a diffraction pattern caused by a large number of randomly placed diffractors. It can also be used to derive numerical predictions for such diffraction patterns. However, it has only limited value for any detailed study of speckles occurring in real rear projector screens. This is due to the fact that it is best suited for treating only point monochromatic sources. Such a model was set up and rapidly analyzed, but we do not describe it here.

2.2 Modeling and Analysis

Since diffraction phenomena cannot be caused by incoherent light sources, the extended light source was modeled by using the theory of partial coherence that is described in Chapter 10 of Born and Wolf [1]. This theory treats an extended source as a union of coherent independent light sources. The mathematical simplification that this theory affords is due to the fact that it reduces the study of light intensities from a large number of point sources to that of the behavior of the correlation Γ_{12} of the electromagnetic field at two points in space labelled one and two. Γ_{12} involves the average of large times of the wave functions, and this theory cannot be used to study fluctuations of these functions with time which, incidentally, cannot be observed. However, it does lead to a straightforward way of predicting intensity distributions and the effects that various parameters have on such distributions.

Scintillations were studied using this model in two ways. First, a computer program was written which predicted intensity distributions across a viewing plane such as the viewing plane in Fig. 2.1. This program was used to numerically study the effects that parameters such as the size of the source have on the scintillation phenomenon. These results are described later in this section.

The second task for which this model was used was that of mathematically predicting overall properties of scintillations on screens where the diffusing effect is due to a large number of randomly placed particles. The methods used here are of a statistical nature and yield a number of predictions that could be experimentally observed. We now take up the description of the model and these results. We first note that there is a complementarity principle in the linear wave theory which allows one to replace a transparent screen coated with opaque diffractors with an opaque screen that has apertures in place of the diffractors.

For a screen which is opaque, expect at a single aperture A, the Fraunhoffer diffraction formula yields the expression

$$U(P) \; = \; C \int\int_A e^{ik(p\xi + q\eta)} d\xi \, d\eta. \tag{2.1}$$

The derivation of this formula can be found in Chapter 8, Sections 2 through 5 of [1]. Here U is the scalar field strength, A is the aperture, (ξ, η) is a Cartesian coordinate system on the plane of the screen, $k = 2\pi/\lambda$, λ being the wavelength of the sources, $p = \ell - \ell_0$, $q = m - m_0$ where ℓ_o, m_0 are the direction cosines in the ξ, η directions from A to P_0, and ℓ, m are direction cosines from A to P_1, and C is a phase and amplitude factor. In the case of many identical holes distributed on the plane of the screen according to some density $\Phi(x, y)$, we let N be the number of holes and write the contribution from the ith hole as $U_i(P)$. The total intensity is given by

$$I(P) \; = \; \left| \sum_{i=1}^{N} U_i(P) \right|^2$$

Here, we let $(x, y) = (x_1, y_1; x_2, y_2; \cdots; x_N, y_N)$ describe the coordinates (x_i, y_i) of the ith hole for $i = 1, 2, ..., N$.

For the extended source in the partial coherence model we assume that the light emitted from a point on the source is statistically independent of the light emitted by any other point on the source. This assumption is valid for all real light sources with the exception of lasers. The notation and coordinate systems for this model are shown in Fig. 2.1. The holes are centered at points (x_i, y_i) on a plane A and the observer is at a plane B described by the coordinates (u, v) as indicated by the axis shown. We assume that the waves propagate in a homogeneous medium with propagation velocity c.

Our goal is to find $I(Q)$, the measurable intensity at a point Q on the observing plane. The intensity is given by the long time average of the complex square of the wave disturbance. In partial coherence theory this is given by the mutual coherence function evaluated at $Q_1 = Q_2 = Q$ and $\tau = 0$, i.e.

$$I(A) = \Gamma^B(Q,Q,O) = < V(Q,t)V^*(Q,t) > \qquad (2.2)$$

where the brackets indicate the time average

$$< f(P,t) > = \lim_{T \to \infty} \frac{1}{2T} \int_{-T}^{T} f(P,s)ds.$$

To find this function after it has passed through the holes to plane B we apply the propagation law for the mutual coherence function

$$\Gamma_B(Q_1,Q_2,\tau) = \int_0^\infty d\nu \int_\sigma V_d(Q_1,t_1,\alpha,\nu)V_d^*(Q_2,t_2,\alpha,\nu)d\alpha,$$

where

$$V_d(Q_i,t_i,\alpha,\nu) = \int_A V(P_i,t_i,\alpha,\nu)\frac{e^{ikr_i}}{r_i}\Lambda(P_i,t_i,\alpha,\nu)dP_i.$$

is the usual diffracted wave from classical diffraction theory. From Eq. (2.2) and (2.3) we get

$$I(Q) = \int_0^\infty d\nu \int_\sigma V_d(Q,\alpha,\nu)V_d^*(Q,\alpha,\nu)d\alpha \qquad (2.3)$$

where

$$V_d(Q,\alpha,\nu) = e^{2\pi\nu t}\int_A \sqrt{I(\alpha,\nu)}\,\frac{e^{iks}}{s}\frac{e^{ikr}}{r}\Lambda(P,Q,\nu)dP$$

and

$$\Lambda(P,Q,\nu) = -\frac{i}{\lambda}\cos\chi,$$

where χ is the angle between the normal to the screen and the vector connecting (x_i,y_i) and the point Q on the viewing plane.

Considering only the integrand of Eq. (2.3) we see that V_d is the classical diffracted wave disturbance from the screen for each point α on a source having intensity $\sqrt{I(\alpha,\nu)}$ and frequency ν. Thus, we see that the partial coherence result can be obtained from the classical diffraction model by integrating the monochromatic point source intensity over all points of the extended source (using the amplitude $\sqrt{I(\alpha,\nu)}$ for each point) and over all frequencies. However, this is not as simple as it appears because of the complexity of the integrand. Here, partial coherence theory provides a simplifying procedure. Since the screen is opaque, we will only have contributions to the diffracted wave V_d at those position where there is a hole. Then, we write the integrand of Eq. (2.3) in the form

$$I(Q,\alpha,\nu) = \sum_{i=1}^N |V_d^{(i)}(Q,\alpha,\nu)|^2 + \sum_{i\neq j}^N \sum_j^N V_d^{(i)}(Q,\alpha,\nu)V_d^{(j)*}(Q,\alpha,\nu), \qquad (2.4)$$

where

$$V_d^{(i)}(Q,\alpha,\nu) = e^{2\pi i\nu t} \int_{\Sigma_i} \sqrt{I(\alpha,\nu)}\, \frac{e^{ik(r+S)}}{rS} \Lambda(\Sigma_i,Q,\nu) d\Sigma_i,$$

and Σ_i denotes the ith hole. We recognize that the first two terms give the intensity diffraction pattern for each hole alone, which we shall denote by $I^{(i)}(Q,\alpha,\nu)$. We now make the assumption that the holes are small enough so that the inclination factor $\Lambda(\Sigma_i,Q,\nu)$ varies very little across the ith hole. Then the second term in Eq. (2.4) can be written as

$$\sum_{i\neq j}^{N} \sum_{j}^{N} V_d^{(i)}(Q,\alpha,\nu) V_d^{(j)*}(Q,\alpha,\nu) =$$

$$\sum_{i\neq j}^{N} \sum_{j}^{N} \Lambda(\Sigma_i,Q,\nu)\Lambda^*(\Sigma_j,Q,\nu)\Gamma^A(\Sigma_i,\Sigma_j,\alpha,\nu,\frac{r_j-r_i}{c})$$

(2.5)

where we have obtained the mutual coherence function per unit source area per unit frequency interval across the plane A. Since the holes are small, the coherence function is evaluated at the centers of holes i and j. However, since the inclination factors are difficult to work with, we rewrite (2.5) in terms of the complex degree of coherence given by

$$\gamma(P_1,P_2) = \frac{\Gamma(P_1,P_2,\tau)}{[I(P_1)I(P_2)]^{\frac{1}{2}}},$$

and use the fact that we can pair the terms as complex conjugates to obtain

$$2\sum_{i<j}^{N-1} \sum_{j}^{N} \left| \Lambda(\Sigma_i,Q,\alpha)I(\Sigma_i,\alpha,\nu)\Lambda^*(\Sigma_j,Q,\alpha)I(\Sigma_j,\alpha,\nu)\right|^{1/2}$$

$$\times \gamma^{(r)}(\Sigma_i,\Sigma_j,\alpha,\nu,\frac{r_j-r_i}{c}),$$

where the superscript (r) denotes the real part. But, the product of the inclination factor at Σ_i with the intensity at Σ_i for each α and ν is, according to diffraction theory, the intensity at Q from the ith hole. This term can be further simplified to

$$2\sum_{i\neq j}^{N-1} \sum_{j}^{N} \left| I^{(i)}(Q,\alpha,\nu)I^{(j)}(Q,\alpha,\nu)\right|^{1/2} \gamma^{(r)}(\Sigma_i,\Sigma_j,\alpha,\nu,\frac{r_j-r_i}{c}).$$

Using the simplifications above and the notation $\gamma^{(r)}(\Sigma_i,\Sigma_j,\alpha,\nu,\frac{r_j-r_i}{c}) = \gamma_{ij}^{(r)}(\alpha,\nu,\frac{r_j-r_i}{c})$, we obtain the monochromatic point source intensity from Eq. (2.4) in the form

$$I(Q,\alpha,\nu) = \sum_{i=1}^{N} I^{(i)}(Q,\alpha,\nu)$$

$$+ 2\sum_{i<j}^{N-1} \sum_{j}^{N} \left| I^{(i)}(Q,\alpha,\nu)I^{(j)}(Q,\alpha,\nu)\right|^{1/2} \gamma_{ij}^{(r)}(\alpha,\nu,\frac{r_j-r_i}{c}).$$

(2.6)

We can then obtain the intensity diffraction pattern for an extended polychromatic source according to Eq. (2.2) by integrating Eq. (2.6) over α and ν.

Since the scintillation phenomenon we are modeling is viewed with a color sensitive device, namely the human eye, we know that we can wash out the speckle effect only by washing out each monochromatic pattern. Thus, we consider only the monochromatic intensity pattern for an extended source by integrating Eq. (2.6)) over σ to yield

$$I(Q,\nu) = \sum_{i=1}^{N} I^{(i)}(Q,\nu) + 2 \sum_{i<j}^{N-1} \sum_{j}^{N} \left| I^{(i)}(Q,\nu) I^{(j)}(Q,\nu) \right|^{1/2} \gamma_{ij}^{(r)}(\nu, \frac{r_j - r_i}{c}),$$

(2.7)

where

$$I^{(i)}(Q,\nu) = \int_{\sigma} I^{(i)}(Q,\alpha,\nu) d\alpha,$$

$$\left| I^{(i)}(Q,\nu) I^{(j)}(Q,\nu) \right|^{1/2} \gamma_{ij}^{(r)}(\nu, \frac{r_j - r_i}{c})$$
$$= \int_{\alpha} \left| I^{(i)}(Q,\alpha,\nu) I^{(j)}(Q,\alpha,\nu) \right|^{1/2} \gamma_{ij}^{(r)}(\alpha, \nu, \frac{r_j - r_i}{c}) d\alpha.$$

In order to simplify the analysis of Eq. (2.7), we make the additional assumption that the source has a uniform intensity, i.e. $I(\alpha,\nu) = I(\nu)$. Then we write $I^{(i)}(Q,\alpha,\nu)$ explicitly as

$$I^{(i)}(Q,\alpha,\nu) = I(\nu) \left| \int_{\Sigma_i} \frac{e^{iks_i}}{S_i} \frac{e^{ikr_i}}{r_i} \Lambda(\Sigma_i, Q, \nu) d\Sigma_i \right|^2,$$

where $k = 2\pi/\lambda$. But since we said the holes were very small, this becomes approximately

$$I^{(i)}(Q,\alpha,\nu) \simeq \frac{I(\nu) \cos^2 \chi}{\lambda^2 r_i^2} A_{\Sigma_i} \frac{1}{s_i^2}$$

where A_{Σ_i} is the area of the ith hole.

We want to compare the above expression for $I^{(i)}$ with the complex degree of coherence which is given by

$$\gamma_{ij}^{(r)}(\alpha, \nu, \frac{r_j - r_i}{c}) = e^{ik(S_i - S_j)} e^{ik(r_j - r_i)}.$$

(2.8)

Thus, we see that if

$$k \gg \frac{1}{S_i - S_j},$$

(2.9)

then $\gamma_{ij}(\alpha, \nu, \frac{r_j - r_i}{c})$ will vary much more rapidly with α than will $I^{(i)}(Q,\alpha,\nu)$. If this is the case then we can approximate (2.6) by letting $I_o^{(i)}(Q,\nu)$ be the intensity diffraction pattern of a point source having intensity $I(\nu)A_\sigma$, where A_σ is the area of the source, located at the center of the source. We do this because we can then use the well known formulas of classical diffraction theory to find $I_o^{(i)}(Q,\nu)$ for certain types of holes. Then the dependence on the size of the source is entirely in the complex degree of coherence $\gamma_{ij}(\nu, \frac{r_j - r_i}{c})$.

Now that we have a formula for the intensity diffraction pattern we consider a special case. We take N identical circular holes each having radius a and a circular source having radius ρ. The monochromatic point source intensity is well known (see, Chapter 8, Section 5 of [1]) and is:

$$I_o^{(i)} = \frac{\pi a^2}{\lambda^2} E_i \text{ circ}^2 \text{ (kaw}_1), \tag{2.10}$$

where $\text{circ}(x) \equiv 2J_1(x)/x$ with J_1 denoting the Bessel function of order one, and

$$s_i = (S^2 + x_i^2 + y_i^2)^{1/2}, \ r_i = (R^2 + (u - x_i)^2 + (v - y_i)^2)^{1/2},$$

$$P_i = \frac{u - x_i}{r_i} + \frac{x_i}{s_i}, \quad q_i = \frac{v - y_i}{r_i} + \frac{y_i}{s_i}, \quad w_i = (p_i^2 + q_i^2)^{1/2}.$$

The calculation of the complex degree of coherence for the extended circular source $\gamma_{ij}(\nu, \frac{r_j - r_i}{c})$ has already been derived by von Laue (see pp. 16-20 of [4]) and it yields

$$\gamma(\nu, \frac{r_j - r_i}{c}) = e^{ik\psi_{ij}} e^{ik(r_j - r_i)} \text{ circ } (k\frac{\rho}{s}\ell_{ij}), \tag{2.11}$$

where

$$\psi_{ij} = \frac{(x_i^2 - y_i^2) + (x_j^2 - y_j^2)}{2S}; \ \ell_{ij} = ((x_i - x_j)^2 + (y_i - y_j)^2)^{1/2}.$$

Combining Eqs. (2.10) and (2.11) in (2.6) gives us the monochromatic intensity diffraction pattern for the "circular" multihole problem with an extended source:

$$I(Q, \nu) = \frac{\pi a^2}{\lambda^2} \Big[\sum_{i=l}^{N} E_i \text{ circ }^2(\text{kaw}_i)$$

$$+ 2 \sum_{i=1}^{N-1} \sum_{j=i+1}^{N} |E_i \text{ circ } (\text{ kaw}_i)E_j \text{ circ } (\text{kaw}_j)| \tag{2.12}$$

$$\times \text{ circ}(k\frac{\rho}{S}\ell_{ij}) \cos\big(k(\psi_{ij} + (r_j - r_i))\big)\Big].$$

This is the formula which can be used to numerically calculated the various scintillation patterns, however, we shall not show the numerical results here since they are difficult to depict without the use of color. Details of the numerical work are in the report [2] and show a scintillation phenomenon which depends on frequency and source radius ρ. From Eq. (2.12) we can mathematically deduce the result that are shown in these numerical calculations and derive a number of other conclusions concerning this phenomenon.

The first term in the intensity Eq. (2.12) is a background intensity which is the sum of all the diffraction patterns from each hole. As we showed earlier, if condition (2.11) is satisfied, then this term is nearly independent of the radius of the source. In the second term, however, the dependence on the source size appears. This second term represents the interference between holes with the degree of interference being governed by the complex degree of coherence. In particular, the factor

$$\operatorname{circ}\left(k\frac{\rho}{S}\ell_{ij}\right)$$

enters in this term

We see that, if the source radius is small, then $\operatorname{circ}(k\frac{\rho}{S}\ell_{ij})$ is large and the oscillatory nature of the interference terms in the intensity formula contribute significantly to the scintillation. If on the other hand ρ is large, then regardless of the other factors in $\operatorname{circ}(k\frac{\rho}{S}\ell_{ij})$ the interference term will be small and not contribute to the intensity, thus reducing the scintillation. We can also vary k, S, or ℓ_{ij} to try to achieve the same effect. However, these terms, or rather functions of these terms, appear in all the contributions to the intensity in formula (2.12); so one must be careful in drawing any conclusions. In particular, by increasing k (decreasing λ), decreasing S, or increasing the values of ℓ_{ij} (spacing the holes further apart) we can decrease or remove the interference terms which vary more rapidly than the diffraction terms. But then, the diffraction patterns of the individual holes become more localized and a different type of pattern will emerge.

We can draw the following conclusions from Eq. (2.12):

1. Increasing the source radius (and in general, the source size for any shaped source) in a spatially incoherent extended source will reduce the intensity fluctuations in a multihole screen.
2. Decreasing the source to screen distance will also reduce intensity fluctuations up to a certain point.
3. Increasing the hole separations in the screen will reduce the interference fluctuations. However, the diffraction patterns of the individual holes will become more apparent.
4. Decreasing the wavelength will decrease the interference effects. This means that we predict that, in a polychromatic pattern, the lower wavelength colors would have less intense speckles.

2.3 The Statistics of Scintillations

Equation (2.12) is good for computer numerical work, but it is difficult to work with it mathematically since it is entirely dependent on where the holes are placed. We shall now proceed to a method that will help us answer questions about the general behavior of a screen of the type that we have been describing. The fact that we are dealing with large numbers of randomly distributed diffraction apertures tells us that we should adopt a probabilistic approach to the problem. We now ask the following question: If we take a large collection of screens having certain characteristics in common then, what will we observe on the average? Averages of this type are called ensemble averages, which we call means for simplicity.

The things which we shall take to be common in the ensemble are the source and its distance from the apertures, the distance of the viewing plane from the aperture plane, the wavelength for each monochromatic case, and the number of apertures. Again we will limit the scope of this analysis to the case of identical holes; thus reducing the number of random parameters to just the location of the holes. What is left to specify is the probability density of $((x_1, y_1), (x_2, y_2), ..., (x_N, y_N))$ of the positions the holes on the screen.

In order to obtain the simplifications we want, we make the assumption that the holes are located independently of each other and with identical distributions $\Phi^{(i)}$. This is not the case in general because the holes depend on each other, at least to the extent that two holes do not intersect. However, for small holes and low density of holes this is not a bad assumption.

The ensemble average, or mean, of a function of the holes shall be denoted by a bar over the function. The mean of f is then given by

$$\overline{f} = \int \int \cdots \int f(x_1, ..., x_N, y_1, ..., y_N) \prod_{i=1}^{N} \Phi^{(i)}(x_i, y_i) dx_i dy_i$$

where we used the fact that the holes are independently distributed. We will also consider the variance, which we will denote by $\overline{V}$, and the correlation, which we denote by $\overline{C}$. These are given in terms of the mean

$$\overline{V}f \equiv \overline{(f - \overline{f})^2} = \overline{f^2} - \overline{f}^2$$

$$\overline{C}fg \equiv \overline{(f - \overline{f})(g - \overline{g})} = \overline{fg} - \overline{f}\ \overline{g}$$

First we consider the mean intensity in Eq. (2.6). Since the $I^{(i)}$ are functions of only one hole, the coherence a function of only two holes, and both of these do not change from hole to hole (since the holes are identical), then the mean intensity reduces to

$$\overline{I}(Q, \nu) = N\overline{I}_1 + N(N-1)\overline{I}_{12} \qquad (2.13)$$

where $I_1 \equiv I^{(1)}(Q, \nu)$ is the diffraction pattern of a single hole, and $I_{12} \equiv (I^{(1)}(Q, \nu)I^{(2)}(Q, \nu))^{1/2}\gamma_{12}^r(\tau)$ is the interference term for two holes. From (2.13) we can see several interesting properties which we state below.

1. The mean intensity for an ensemble of screens can be found from the mean diffraction pattern for a single hole placed according to the density Φ_1 and from the interference pattern for two holes each placed according to Φ_1.

2. If the light reaching the screen is totally incoherent ($|\gamma_{12}| = 0$) then the mean intensity at any point is proportional to the number of holes.

3. If the light is not completely incoherent then the mean interference pattern will be accentuated over the mean diffraction pattern by a ratio of N as N gets large.

The first statement indicates the simplicity of the resulting pattern which we could not have expected by looking at Eq. (2.6) or the complicated pattern it gives. The second statement assures us that what we have done is correct since this is the prediction made by the classical theory for incoherent light. The third statement is the most interesting because it indicates that there are two types of patterns we could observe, one where N is small and the diffraction pattern dominates, and one where N is large and the interference effects dominate. That the interference term grows as N^2 follows from the fact that the points on the mean interference pattern which are maxima or minima are formed by pairs of holes. Thus, at the maxima, pairs of holes tend to constructively interfere increasing the intensity by more than the contribution of a single hole, and similarly at the minima. Before we can say more about (2.13) we will have to know more about the means $\overline{I}_1$ and $\overline{I}_{12}$, which means we have to specify the hole density and type of holes more completely. We next consider three other functions which we can average to determine something more about the diffraction pattern than what the mean intensity gives.

The obvious question to ask about the mean intensity is how well does it represent an individual screen. The answer is given by the variance of the intensity for our ensemble of screens. Using the definition of the variance given earlier, and the simplifying conditions leading to Eq. (2.13) we obtain

$$
\overline{V}I(Q,\nu) = N\overline{V}I_1 + 2N(N-1)\overline{V}I_{12} + 4N(N-1)\overline{C}I_1 I_{12}
$$
$$
+ 4N(N-1)(N-2)\overline{C}I_{12}I_{23}. \tag{2.14}
$$

As with the mean we can say several interesting properties about the variance.

1. The variance is a linear combination of the variance of a single hole, the variance of the interference term for a pair of holes, the correlation of the intensity from a single hole with the interference term for that hole with another, and the correlation of the interference terms for three holes taken two at a time, i.e. the correlation between the interference term of holes one and two with that of holes two and three.

2. If the light reaching the screen is totally incoherent, then the variance is proportional to the number of holes, the proportionality constant being the variance of a single hole.

3. If the light is not completely incoherent, then the terms which are due to the correlation between the interference of pairs of holes will be accentuated over the terms which are functions of a single hole and the terms which are functions of a pair of holes in the ratio of N^2 and N, respectively.

Similar statements to those about the mean intensity can be made about the variance. There is a difference in the third statement which is interesting. There are three levels of variance: for N small, N intermediate and N large, with the point at which the intermediate behavior holds depending on the degree of coherence of the light. Combining this with the results for the mean intensity indicates very strongly that there are at least two types of diffraction patterns depending on whether N is small or large.

The next probabilistic function we consider is the correlation which is a generalization of the variance. This involves comparing deviations of the intensity at two points Q_1 and Q_2 from the mean intensity at these points and then taking an ensemble average. Let I be the intensity at Q_1 and J the intensity at Q_2. Carrying out the necessary operations on Eq. (2.6) and using our assumptions yields the correlation

$$
\overline{C}I(Q_1,\nu)J(Q_2,\nu) = N\overline{C}I_1 J_1 + 2N(N-1)\overline{C}I_{12}J_{12}
$$
$$
+ 2N(N-1)(\overline{C}I_1 J_{12} + \overline{C}I_{12}J_1) + 4N(N-1)(N-2)\overline{C}I_{12}J_{23}. \tag{2.15}
$$

This result is of the same form as (2.14) and, in fact, is identical to it when $Q_1 = Q_2$. This function gives us average properties of the detailed structure on the viewing plane, unlike the mean intensity and variance which give an envelope of the structure and its variance.

The correlation can give the average size of features on the viewing plane in the following manner. First, fix a point Q_1 on the viewing plane. Then by the definition of the correlation function, when $Q_1 = Q_2$, Eq. (2.15) will yield a local maximum (which says the points Q_1 and Q_2 are correlated). As we move Q_2 away from Q_1 this function will decrease as the intensities at the two points will, on the average, differ. We then look for the point at which the correlation again starts to increase. At this point we have an average distance from Q_1 for which the intensities differ by the greatest amount in a region about Q_1. This would be the average size of intensity peaks or valleys which is the scintillation phenomenon we wish to study.

Before we go into the last technique we should mention that, like the variance, the correlation indicates three types of behavior in the diffraction pattern depending on the number of holes.

There is one last method by which we can tell something about the finer structure of the diffraction patterns. This involves picking a point Q on the observing plane around which we wish to find the behavior of the intensity. Then for each member of the ensemble of screens we construct a square having sides $2z$ about the point Q over which we calculate the variance per unit area. Then we take the ensemble average of the result.

If we describe the viewing plane by the cartesian coordinates (u, v), and Q by the coordinates (u_0, v_0) then the mean of a function f calculated over the square at Q is given by

$$E_{uv}f = \left(\frac{1}{2z}\right)^2 \int_{u_0-z}^{u_0+z} \int_{v_0-z}^{v_0+z} f(u,v)dudv,$$

where E refers to "expected value" and the subscripts indicate that it is over the observing plane. Then the quantity we want to study is

$$\overline{V_{uv}I(Q,\nu,z)} = \overline{E_{uv}(I - E_{uv}I)^2}.$$

Using the simplifications of identical and identically distributed holes we obtain

$$\overline{V_{uv}I(Q,\nu,z)} = N\overline{V_{uv}I_1} + 2N(N-1)\overline{V_{uv}I_{12}} + 4N(N-1)\overline{C_{uv}I_1 I_{12}}$$
$$+ 4(N)(N-1)(N-2)\overline{C_{uv}I_{12}I_{23}}, \tag{2.16}$$

which is not surprising in light of the previous results. The same observations again apply to Eq. (2.16) as applied to Eqs. (2.13) and (2.15).

The reason behind using a function of this type is somewhat subtle. Suppose the intensity across the observing plane for each screen consists of many intensity peaks (scintillations) spaced some average distance apart. As we start with z very small the variance must be nearly constant in the square. Upon increasing z the variance will rise at a rate depending on whether Q is on the edge of speckle, the center, or in between two speckles. Each time another peak passes into the square, the variance should rise steeply. Finally, when the square encloses the entire pattern (since the number of holes is finite this must always occur), the variance will go down, since it is computed per unit area, and eventually reach zero when the area becomes infinite. The subtlest point, which was not discovered until a numerical example was tried, is that the variance does not rise monotonically even when the square is within the pattern. This is because if the scintillations are far enough apart, then it is possible for the amount of area without intensity peaks to grow fast enough so that the effect of the scintillations diminishes and the variance goes down until the next peak is reached. Thus, from the above we see that what we want to do is find the first point at which the variance increases rapidly as z increases from zero. This point will mark a region about Q before an edge of a scintillation is reached, either because we are on top of a peak or because we are in between scintillations. This type of analysis can lead to predictions of the size of speckles and hence make possible a study of the effects that various parameters can have on the scintillations.

With these probabilistic formulas we can analyze scintillation sizes and distributions. We conclude that there are two (and possibly three) types of patterns which can occur, depending on the number of holes. The case of large N corresponds to realistic models of real projection screens and can be used to study the scintillation found in such rear projection screens.

2.4 Results and Conclusions

The mathematical analysis and the resulting methods for computing the intensities and distribution of scintillations was of interest to the Bell and Howell scientists. The managers, however, were more interested in the possibility of using the insights provided by this theoretical work in order to eliminate, or at least to improve, the scintillations appearing on the screens. Consequently, the mathematical and numerical results needed to be interpreted in language that did not require a very technical physical or mathematical background. Here I will collect the main part of the conclusions and results which were presented in this format.

The scintillation phenomenon is an electromagnetic diffraction effect that is due to a large number of randomly distributed diffracting particles. The first thorough investigations of this effect date back to the early years of this century and are connected with such illustrious names in physics as C. V. Raman [3] and M. von Laue [4, 5]. Work on this type of phenomenon seems to have ceased after these early investigations until the advent of laser light and the observation of laser speckles. A number of studies of laser speckles have been made. However, the fact that laser speckles and screen scintillations are similar phenomena is not widely recognized, and this project was among the first to draw attention to it. All of this previous work, as well as the present models, show that scintillation can be adequately described as an electromagnetic diffraction effect due to large numbers of randomly distributed diffracting particles. This project has extended these investigations and used the theory of partial coherence to construct and analyze a model which give a deeper understanding of the undesirable scintillation phenomenon. We list a number of possible means of reducing the scintillations that are suggested by the analysis of our mathematical models.

The first method of reducing scintillations involves increasing the size of the light source. This can be done by having a bigger aperture for the projection lens, by providing for a method of mechanically moving the light source with enough rapidity so that the eye would average the intensity from the various locations of the source without perceiving the motion, and finally by constructing a double screen with the first screen essentially acting as an enlarged light source that is close to the viewing surface. This type of screen has been patented by Douglas B. Huber. The effect is, of course, a radical reduction in scintillation. This result can be readily predicted from the partial coherence theory model.

Another conceptually simple method of reducing scintillation involves using a screen where the diffracting particles are in colloidal suspension and thus undergoing random Brownian motions. The averaging properties of the human eye would tend to integrate the rapidly varying speckle patterns on such screens and thus reduce the amount of scintillation. A mechanical vibration of any diffusing screen would have the same effect but may be more difficult to attain in an actual physical device. It should be noted that this "washout" of speckles can be easily observed by inducing rapid oscillations in a screen illuminated by either laser or partially coherent light.

Yet another way of reducing the extent of scintillation is by designing the optics of the projection system so that the light source is effectively as close to the screen as possible. The predictions of the partial coherence theory model show that this will tend to reduce scintillation. This is another reason why the screen invented by D. Huber is effective in reducing speckles. The same effect can be achieved by other optical means of bringing the light source closer to the screen.

Finally, it should be noted again that scintillations are a diffraction phenomenon that cannot be totally eliminated when diffusing screens are used. All of the above suggestions provide means of either reducing the scintillations to the point that their nuisance effect is negligible or else of presenting to the eye (or to any other averaging optical device) a pattern that varies so rapidly that its average does not show scintillations or motion. For stationery diffusing screens the number of speckles will decrease as the size of the diffracting particles increases, but the quality of the image on such screens suffers due to the coarseness of the particles, and this is not a viable method of reducing scintillations. On the other hand, even though the number of speckles increases on screens with a number of diffracting particles, such screens are often desirable due to other imaging characteristics. It is hence safe to recommend that a screen should be selected due to its other desirable imaging properties and that the scintillation phenomenon be then reduced by using one or more of the methods described in the previous paragraphs.

Perhaps it is worth concluding with a note on one aspect of this industrial mathematics work which is peculiar to it and which makes it different from usual academic research, namely, the requirement for some confidentiality. The value to the client of the work on the project described above was sufficient that they requested that it not be published or otherwise divulged for one year after the completion of the project. The agreements on the funding of Mathematics Clinics include a clause that gives the client this option and which is designed to allow the company to apply for patents or take other steps which would give them an advantage in commercially exploiting the results of the projects they sponsor. This clause is exercised by the clients in a substantial number of the projects. After this one year period, the team members have can use the work in their reports as a basis for writing professional papers or for presentations in scientific meetings. Although this may present some difficulties, particularly because a number of the students who work on the projects often go on to other pursuits after the completion of their studies, it also provides the positive aspect of allowing a more leisurely and reflective approach to the decision of what should and should not be published.

References

1. Born, M. and E. Wolf, (1970): Principles of Optics, fourth edition. Pergamon Press, New York
2. Busenberg, S., S. Quist, and F. Valdes, (1973): Analysis of Scintillation Phenomena in Rear Projection Screens, Harvey Mudd College Mathematics Clinic Report #1.
3. Raman, C. V., (1919): *Phil. Mag.*, p. 568-.
4. Von Laue, M., (1914): *Sber. Preuss Akad. Wiss.*, (English translation, California Institute of Technology), **47**, p. 1144-.
5. Von Laue, M., (1919): *Mittt. Phys. Ges. Zürich*, **18**, p. 90-.

3 Identification of Semiconductor Properties

3.1 Description of the Problem

VLSI (Very Large Scale Integrated) semiconductor devices are commonly used in a multitude of components and machines ranging from such everyday appliances as radios to sophisticated guidance systems in rockets and aircraft. There has been a continuous quest for reducing the size of these VLSI devices which has led to vast improvements in the design of the components in which they are used. It is not uncommon these days to find VLSI devices that measure less than one centimeter on the side and which contain hundreds of transistors whose electrical interconnections on the device cannot be seen without the use of a microscope. Together with the advantages of such miniaturization have also come some serious problems in measuring the electrical properties of their tiny contacts, in dissipating the heat that is generated by the currents that pass through them, in controlling the strength of the electrical contacts, and in the overall statistical quality control of their production. The Mathematics Clinics have done several projects in each one of these problem areas, and we shall present one example that has come out of these projects.

In an integrated semiconductor component, a contact is a region where the semiconductor is attached to a metal layer which can be used to connect the semiconductor to external voltage or current sources and drains. The electrical resistance at the contact interface determines the effectiveness of the device since it acts as an undesirable heat source. This resistance is measured via the resistivity ρ_c which becomes a dominant design factor as components are reduced to submicron size, so it is important to obtain accurate values of ρ_c. Because of the small size of these devices, it is impossible to directly measure the contact resistivity and to exactly control the location of the contact window. In order to estimate ρ_c and detect the location of the contact window, current is applied to the component and voltage is measured at some accessible point away from the contact region. The literature contains extensive experimental and computer simulations on this identification problem (see [4] and its references, for example.) The problem is to find a reliable method to obtain the contact resistivity from available measurements and to compare it with heuristic methods that are widely used for estimating ρ_c.

This problem arising from the Clinics was further analyzed in the Ph.D. thesis [3] of Weifu Fang at the Claremont Graduate School, and in the joint work [1] by Fang and myself. The description that we give below is based on these two references, and it provides one example of the type of work that is directly inspired by Clinic projects. In turn, such work increases the expertise of the mathematics faculty in areas which are of current interest to industry. This fact, when coupled with the personal contacts between industrial scientists and academic mathematicians which are part of the structure of Clinic projects, greatly accelerates the transfer of new mathematical methods and techniques from the university to industry.

3.2 The Model and Analysis

Considerable modeling of VLSI devices has been done and one can go to the standard literature for the comparison of the validity of various models. In particular, in the setting in which this problem was posed, the drift diffusion equations are an accepted model for the process. For most applications where small voltages and currents are involved, the small sizes of the VLSI devices lead to a very fast decay (of the other of 10^{-13} seconds) of transients and a rapid settling to a steady state. Consequently, a steady state linear model is appropriate in these situations. Such a model was developed in a Clinic project and is described in [2-3] It leads to the following problem:

The voltage $u(x)$ at position $x \in \Omega$ is approximated by the solution to the elliptic problem

$$\Delta u - p\chi(S)u = 0 \text{ in } \Omega \subset R^2 \tag{3.1}$$

$$\frac{\partial u}{\partial n} = g \geq 0 \text{ but } g \not\equiv 0 \text{ on } \partial\Omega \tag{3.2}$$

where $S \subset \Omega$ and $p > 0$ are unknown, and $\chi(S)$ is the set characteristic function of S. The known density of the applied current is g, S is the location of the contact region, and p is a given positive function of the contact resistivity. There are standard methods for measuring the sheet resistance R_s of the semiconductor layer, hence, it can be assumed to be known. In the above equations $p = R_s/\rho_c$, where ρ_c is the unknown that needs to be retrieved.

The problem then reduces to recovering p and S from a one-point boundary measurement $u(x_0)$ for some $x_0 \in \partial\Omega$. We shall here describe the problem of identifying p when the contact location S is known.

3.3 The Extracted Contact Resistivity

In applications the voltage u is measured at some $x_0 \in \partial\Omega$ and the following formula is used to yield an approximation of the contact resistivity

$$\rho_{ce} = |S| \frac{u(x_0)}{R_s} \int_{\partial\Omega} g \, ds. \tag{3.3}$$

This formula is based on a simple argument using Ohm's law (see e.g. [4-5] for details). Computer simulations of this problem are also given in [5]. In this section we study the quantitative behavior of this extracted contact resistivity ρ_{ce} in terms of the true resistivity ρ_c. Let Ω be a bounded connected domain in R^2 and S a given subdomain in Ω with $C^{1,1}$ boundaries $\partial\Omega$ and ∂S. Points in R^2 are denoted by $x = (x_1, x_2)$. Suppose $g(x) \in C^\alpha(\partial\Omega)$ for some $0 < \alpha < 1$ and $g \geq 0$ but $g \not\equiv 0$.

It is well known that the $H^1(\Omega)$ solution to (3.1), (3.2) is in $C^{1,\beta}(\Omega)$ for some $0 < \beta < 1$. Denote this solution by $u(x; p)$. Our identification problem is: recover the positive constant p from the one-point boundary measurement $u(x_0)$, given the geometric setting and the applied current g on the boundary. The following result provides the basis for the computation of p.

Theorem 3.1. *The solution $u(x;p)$ to (3.1), (3.2) is C^∞ in p, and the kth order derivative with respect to p, denoted by $u^{(k)}(x;p)$, is the solution to*

$$\Delta u^{(k)} - p\chi(S)u^{(k)} = k\chi(S)u^{(k-1)} \quad \text{in } \Omega$$

$$\frac{\partial u^{(k)}}{\partial n} = 0 \quad \text{on } \partial\Omega \qquad (3.4)_k$$

where $k = 1, 2, \cdots$ and $u^{(0)} = u$. Moreover, we have

$$(-1)^k u^{(k)}(x;p) \geq 0 \quad \text{in } \bar\Omega \quad \text{and} \quad (-1)^k u^{(k)}(x;p) > 0 \quad \text{in } \bar\Omega \setminus S. \qquad (3.5)_k$$

The proof of this result is based on standard elliptic estimates and can be found in the paper [1]. Without loss of generality, we study the quantitative behavior of the extracted contact resistivity given by (3.3) for the case where $R_S = 1$, i.e. $\rho_c = 1/p$. In this situation, the extracted contact resistivity ρ_{ce} is a function of the true ρ_c given by

$$\rho_{ce} = f(\rho_c) \equiv \frac{|S|}{\int_{\partial\Omega} g\,ds} u\left(x_0; \frac{1}{\rho_c}\right) \qquad (3.6)$$

where $u(x;p)$ is the solution to (3.1), (3.2) and x_0 is a point at $\partial\Omega$ where the measurement is made. For the so-called Kelvin resistor (a special choice of the density distribution g and the measurement location x_0), some asymptotic properties of ρ_{ce} are observed in [5] by using computer simulations. In general, we have the following properties for f which agree with the observations made in [5]. Let

$$A_0 = \frac{\int_{\partial\Omega} g\,ds}{|S|} > 0.$$

Theorem 3.2. *The function f given by (3.6) has the following properties:*
(i) f is in $C^\infty(0,\infty)$ and is strictly increasing in $(0,\infty)$;
(ii) As $\rho_c \to \infty$,

$$\frac{f(\rho_c)}{\rho_c} = 1 + O\left(\frac{1}{\rho_c}\right);$$

(iii)

$$\lim_{\rho_c \to 0^+} f(\rho_c) = f_0 = \frac{|S|}{\int_{\partial\Omega} g\,ds} u_0(x_0) > 0$$

where $u_0(x)$ is the solution to

$$\begin{cases} \Delta u_0 = 0 & \text{in } \Omega \setminus \bar{S}, \\ u_0 = 0 & \text{on } \partial S, \\ \partial u_0/\partial n = g & \text{on } \partial\Omega. \end{cases} \qquad (3.7)$$

For the proof of this result we again refer the reader to [1]. From (ii) of this theorem, we can see that this extracted resistivity is a good estimate when the true resistivity ρ_c is large. However, from (iii) it is seen that for small ρ_c the extracted resistivity overestimates ρ_c severely, since the limiting value f_0 of ρ_{ce} as $\rho_c \to 0$ in (iii) is a positive constant. Also (ii) provides a method to calculate f_0. A discussion of the physical interpretations of these properties can be found in [5].

3.4 Identifying the Contact Resistivity.

A numerical iteration scheme for identification of the contact resistivity ρ_c, or equivalently, the constant p, from the one-point boundary measurement can be constructed based on the results of the previous section. Assume the geometric settings of the problem (3.1), (3.2) are given, i.e. Ω, the contact window S, the point $x_0 \in \partial\Omega$ where the measurement is made, and the applied current density g are given. Then we wish to estimate the constant p^* from the measurement u^*. From Theorem 3.1, $u(x_0; p)$ is strict decreasing in p and is convex, so for each measurement $u^* > u_0(x_0)$ ($u_0(x)$ is given by (3.7)) there is a unique $p^* > 0$ such that $u^* = u(x_0; p^*)$ where $u(x; p)$ is the solution to (3.1), (3.2). To ensure unique identifiability, in the following we assume that $u^* > u_0(x_0)$.

By Newton's method of finding a zero of a function, we construct the following iteration scheme:

$$p_{j+1} = p_j - \frac{u(x_0; p_j) - u^*}{u^{(1)}(x_0; p_j)} \qquad (3.8)_j$$

where $u(x; p_j)$ is the solution to (3.1), (3.2) with $p = p_j$, and $u^{(1)}(x; p_j)$ is the derivative of $u(x; p)$ with respect to p given by $(3.4)_1$ with $p = p_j$. Besides the quadratic convergence property of the general Newton's method, the following result can be proved [1] for this specific problem.

Theorem 3.3. *If the initial guess $p_0 > 0$ is such that $u(x_0; p_0) > u^*$, then the sequence $\{p_j\}_0^\infty$ given by Eq. (3.8) is strictly increasing and converges from below to p^*, the unique value such that $u^* = u(x_0; p^*)$.*

We need to consider the appropriate choice of the initial guess p_0. We call p_0 an eligible initial value if $u(x_0; p_0) \geq u^*$, or equivalently, $p_0 \leq p^*$. So if p_0 is eligible, by Theorem 3.3, the Newton iteration scheme (3.8) starting at this p_0 is monotone increasing and converges to p^*. On the other hand, if p_0 is not eligible, i.e. $u(x_0; p_0) < u^*$, then $p_0 > p^*$, and clearly the next iteration p_1 by (3.8) may become nonpositive, so the iteration cannot proceed. In this case, instead of using (3.8) to obtain the next iteration, we set $p_1 = p_0/4$. We can continue this procedure until we come up with a p_i such that $u(x_0; p_i) > u^*$, then we turn to the Newton scheme (3.8) for the rest of the iterations, using this p_i as the initial value. At each iteration step, two elliptic problems, one for u and one for $u^{(1)}$, need to be solved to obtain the next iteration value.

We remark that we can also use the bisection method to identify p^*. For this method, we need to search for the interval on which we start the bisection procedure. Similar ideas as in the search for the eligible initial value p_0 above can be used to find this interval. At each iteration step, only one elliptic problem (for u itself) needs to be solved. Comparing these two schemes, we notice that we compensate the speed of convergence in the bisection method for the ease of solving only one elliptic problem.

Finally, we remark that in the above we assume that there is no noise in the measurement. In the case that there is noise in the measurement but the experiment is repeated for a number of times, we should take the mean value for these data first and then use the above scheme to find the p corresponding to this mean value as the estimate for the true p.

We consider a particular example where Ω and S are two non-concentric discs. Let

$$\Omega = \text{the unit disc}, \quad S = \text{ a disc centered at } (0, 0.2) \text{ with radius } \frac{1}{2}.$$

The applied current density is given by

$$g(x) = \begin{cases} |x_2 + \frac{1}{2}| & \text{when } x_2 \leq -\frac{1}{2}, \\ 0 & \text{when } x_2 > -\frac{1}{2}. \end{cases}$$

And the measurement is made at $x_0 = (1, 0)$.

For any given u^*, we pick an arbitrary initial p_0. Then we solve Eqs. (3.1), (3.2) with this p_0 and test if this p_0 is eligible. If not we reduce to a quarter of p_0 and test again, until we get an eligible initial value. Once we get an eligible p_0, we use the Newton iteration scheme (3.8) to obtain the next value, until the present value is close enough to the previous one, or the calculated u value is close enough to the measured u^*. At each iteration step, we use the elliptic problem solver ELLPACK (see [6]) to solve for $u(x; p_j)$ and $u^{(1)}(x; p_j)$. Notice that in the problem for $u^{(1)}(x; p_j)$, that is (3.4)$_1$, we need the values of $u(x; p_j)$. For comparison we use the bisection method and note that taking initial values for both iteration schemes to be $u_0 = 1.0$ and $p_0 = 2.0$, the Newton method scheme converges to the correct values $p^* = 0.971166$, $u^* = 1.0$ with a six place accuracy in 8 iterations (that is we have to use the elliptic solver 16 times). On the other hand, the bisection method takes 21 steps (requiring us to solve 21 elliptic problems) to arrive to the same accuracy. This method provides an effective way of computing the contact resistivity from standard measurements at the boundary of the VLSI device.

3.5 Conclusions

The method outlined above for extracting the contact resistance from boundary measurements provides both an easily implemented algorithm for computing ρ_c and an analysis of the range of validity of a standard heuristic method that has been used for this purpose. The heuristic method is shown to be valid only when the contact resistance is large, and to severely overestimate the value of ρ_c when it is small, in fact, when it is in the range that occurs in a number of VLSI chips. The work that was described here represents one of the investigations that were stimulated by the projects involving semiconductors that the Mathematics

Clinics have done and continue to do. This is an area where there is a continuing exchange between industry and academia and where the Mathematics Clinics are acting as a catalyst that speeds this transfer of ideas.

References

1. Busenberg, S. and W. Fang, (1991): *Identification of semiconductor contact resistivity*, Quart. Appl. Math., **64**, 639-649.
2. Fang, W. and E. Cumberbatch, (1990): *Inverse problems for MOSFET contact resistivity*, SIAM J. Applied Math., to appear.
3. Fang, W., (1990): *Identification of transistor contact resistivity*, Ph.D. Thesis, Claremont Graduate School.
4. Loh, W. H.,(1987): *Modelling and measurement of contact resistance*, Stanford Electronic Labs., Tech. Rep., No. G830-1.
5. Loh, W. H., K. Saraswat and R. W. Dutton, (1985): *Analysis and scaling of Kelvin resistors for extraction of specific contact resistivity*, IEEE Electron Device Letters, **6**, 105-108.
6. Rice J., and R. F. Boisvert, (1985): *Solving elliptic problems using ELLPACK*, Springer-Verlag, New York.

4 Pattern Recognition with Neural Networks

4.1 Description of the Problem

The problems which were described in the previous two sections dealt with the modeling and analysis of physical phenomena. These types of projects represent about forty percent of the problems that have been studied by the Mathematics Clinics. The remainder of the projects involve problems from either the decision sciences or computer science. We proceed to describe a project [1] which falls in this last category and whose problem area was the source of three clinic projects which were sponsored by the Pomona division of General Dynamics. The statement of the problem was very vague:

> Analyze the feasibility of developing a Neural Network that could automatically recognize images and would retain this ability when the objects are rotated. This image recognition should, when implemented on current microprocessor technology, be fast enough to be done in real time, and be able to adapt to recognizing new images. When this problem was brought to the Clinic over five years ago, the area of neural networks was not as widely developed and popular as it is now. The statement of the problem had to be further developed after several meetings with the industrial liaison, Dr. Richard Schlunt, who had substantial previous experience in nonlinear filtering and in the use of traditional artificial intelligence methods for image recognition. The results of these discussions lead to the understanding that by "neural network" was meant a computer program that was not based on expert rules and standard von Neumann sequential stored instruction programming. Rather, the program was expected to adapt under the influence of external inputs. The images that were to be recognized were two dimensional, and once the neural network "learned" to recognize an image in some standard orientation, then it would still recognize it when it was rotated by up to 30 degrees.

The resulting neural network should after a certain training sequence, recognize future inputs. Since the program was to adapt to external stimuli, "success" could be defined in a few different ways:

1. The learning rule would continue to function after the training session ended, but the system would recognize all future presentations of the training patterns, and possibly learn some new ones.

2. The system would recognize each of the input patterns used in training it. Then the learning rule would be turned off, and it would always recognize these patterns from then on.

3. The learning rule would continue to function, and a reliability condition would be imposed so that not too many patterns would be forgotten too quickly under proper conditions.

The model was designed around the first definition, primarily because it was the most acceptable to the client.

The problem as stated and refined was not yet phrased in any traditional mathematical format that would yield to analysis. Consequently, several approaches were initiated to model the problem and to define in mathematical terms the desired properties imposed by the client. We shall describe here the approach that the client eventually found to be most useful and which became the basis for a microprocessor design.

4.2 Modeling and Analysis

The model equations that were chosen as the basis for the analysis are commonly called Hopfield's discrete neural network. The basic components of this model are as follows. At each discrete point in time t_i, the vector valued function $u(t_i)$ denotes the potential of the neurons of the network with each component $u_j(t_i)$ representing the potential of neuron j at time t_i. The state of these neurons is governed by the discrete dynamical system which gives the potential of each neuron at time $t + 1$ given an external input I.

$$u_j(t+1) = \sum_{j=1}^{N} T_{ij} f_j(u_j(t)) + I_i(t). \tag{4.1}$$

The real matrix T with entries T_{ij}, is called the connectivity matrix and is assumed to be constant. The functions f_j are real valued and are called the neuron response functions. The typical form of the functions f_i is sigmoid, or a soft switching function. Of course, Eq. (4.1) is nothing more than a special form of a nonlinear discrete dynamical system, however, the connection to biological neural networks, nebulous though it may be, yields dividends that go far beyond the enrichment of the terminology. The main benefit of the biological analogy is that particular structures are suggested for the form of the functions and parameters entering in this equation. Essentially, the view is of a large numbto a nonlinear law to the weighted sum of the inputs that they receive. Also, all of this response occurs in parallel in the sense that the ith neuron does not wait for the $(i-1)^{th}$ to finish updating its potential before it updates it own. This is, of course, a familiar banality in the theory of dynamical systems, but is dramatically different from the way that traditional computers are designed.

Adopting this model for the neural network that would analyze the images which are represented by the inputs I, one needs to next decide on the definition of the requirement that an input be recognized. The definition that was adopted is the following. Suppose that when a particular fixed input $\overline{I}$ is given in the neural network Eq. (4.1), the resulting dynamical system has a unique globally attracting steady-state solution $\overline{u}$. Then the event of having the solution reach a sufficiently small neighborhood of this steady state is interpreted as the recognition of the presented image $\overline{I}$. Clearly, rotational invariance would signify that the inputs which correspond to the rotated image would lead to steady states that lie in this neighborhood of $\overline{I}$. Of course, two objects that need to be distinguished would have to result in steady states that are sufficiently separated that their identification neighborhoods do not overlap. This is the basic motivation for the model that we shall present.

We will use the following notation throughout this analysis. The letter x will denote the vector whose entries are x_i. Thus, f will be the column vector whose entries are $f_1, f_2, ..., f_n$. Since the recognition of an object by the network is interpreted as the settling to a unique steady-state, we need to establish conditions under which our model would be guaranteed to have a unique equilibrium which is asymptotically stable. To study the stability of the equilibrium we let $\bar{u}_i$ be a fixed point of (4.1) and consider its linearization about $\bar{u}$,

$$w_i(t+1) = \sum_{i=j} T_{ij} \frac{\partial f_j}{\partial u_j}(\bar{u}_j) w_j(t).$$

Then the steady state of the non-linear system is locally asymptotically stable when the zero solution of this linear system is stable. The matrix for this linearized system is

$$A = \begin{pmatrix} T_{11}\frac{\partial f_1(\bar{u}_1)}{\partial u_1} & T_{12}\frac{\partial f_2(\bar{u}_2)}{\partial u_2} & \cdots & T_{1N}\frac{\partial f_N(\bar{u}_N)}{\partial u_N} \\ T_{21}\frac{\partial f_1(\bar{u}_1)}{\partial u_1} & T_{22}\frac{\partial F_2(\bar{u}_2)}{\partial u_2} & \cdots & T_{2N}\frac{\partial f_N(\bar{u}_N)}{\partial u_N} \\ \vdots & \vdots & \ddots & \vdots \\ T_{N1}\frac{\partial f_1(\bar{u}_1)}{\partial u_1} & T_{N2}\frac{\partial f_2(\bar{u}_2)}{\partial u_2} & \cdots & T_{NN}\frac{\partial f_N(\bar{u}_N)}{\partial u_N} \end{pmatrix}.$$

If the eigenvalues of A lie within the unit circle, then we will have stability, thus, $||A|| < 1$ is sufficient for stability. For a sigmoid-shaped $f_i(u_i)$, we can define threshold points θ_i^1 and θ_i^2 such that $V_1 = f_i(u_i) \approx 0$ for $u_i < \theta_i^1$ and $V_i \approx 1$ for $u_i > \theta_i^2$. When the $f_i' \approx 0$, which occurs for large and small values of $\bar{u}_i$, corresponding to $V_i \approx 1$ and $V_i \approx 0$, respectively, the spectral radius would be expected to be less than one, and the equilibrium would be locally asymptotically stable. Just how "large" or "small" the $\bar{u}_i$ must be is determined by the shape of the f_i and by how large the T_{ij} are. This heuristic reasoning suggests that one should consider the firing rate function $f(u)$ rather than the state u as the basic variable.

Assume $f(u)$ is a steady state under a constant input $I(t) = I$ for all t. Then $f(u(t)) = f(u)$ for all t, and Eq. (4.1) gives

$$f(u) = f(Tf(u) + I).$$

Let $W = f(u)$, and $\phi(W) = f(TW + I)$. Then we are interested in solving the equation $W = \phi(W)$.

Define

$$M = \max_{\substack{1 \le i \le N \\ -\infty \le u_i \le \infty}} |f_i'(u_i)|, \quad T = \max_{1 \le j \le N} \sum_i |T_{ij}|, \quad ||W|| = \max_{\substack{1 \le i \le N \\ -\infty \le u_i \le \infty}} |f_i(u_i)|.$$

The following theorem gives sufficient conditions for three of the desirable properties of the type of neural network that we wish to design, that is, the uniqueness of steady states for a given input, their global stability, and a large lower bound on the total number of distinct steady states.

Theorem 4.1. *For a discrete Hopfield model (4.1) with a smooth, continuous firing rate function f, satisfying*
  (i) f is a monotone increasing function with $\lim_{u_i \to -\infty} f_i(u_i) = 0$ and $\lim_{u_i \to \infty} f_i(u_i) = 1$ for each $i = 1, 2, ..., N$, and
  (ii) $\alpha = \mathbf{MT} < 1$,
for any given fixed input vector I there is a unique steady state response W. Furthermore, letting $W(t)$ denote the response at an integer value of the time t, then

$$\|W - W(t)\| \leq \frac{\alpha^t}{1 - \alpha} \|W - W(0)\| \tag{4.2}$$

If f satisfies
  (iii) there exist thresholds θ_1 and θ_2, $\theta_1 < \theta_2$, such that $f_i(u_i) = 0$ if $u_i \leq \theta_1$, and $f_i(u_i) = 1$ if $u_i \geq \theta_2$,
then the system has at least 2^N distinct steady state responses, where N is the number of neurons.

Proof. Let $S = \{W : 0 \leq W_i \leq 1\}$. By hypothesis (i), $0 \leq \phi_i(W_i) \leq 1$; thus S is a closed and bounded set which is invariant under ϕ. That is, $\phi(S) \subset S$. Next, we will show that ϕ is a contraction on S by showing that

$$\|\phi(W) - \phi(Y)\| \leq \alpha \|W - Y\|,$$

for all $W, Y \in S$. Note that $\phi(W) = (\phi(W_1), \phi_2(W_2), ..., \phi_N(W_N))$, where $W = (W_1, ... W_N)$. Now,

$$\phi_i(W_i) = f_i(\sum_j T_{ij} W_j + I_i).$$

Suppose $W, Y \in S$. Then, for each $i = 1, 2, ..., N$,

$$\phi_i(W_i) - \phi_i(Y_i) = f_i(\sum_j T_{ij} W_j + I_i) - f_i(\sum_j T_{ij} Y_i + I_i).$$

By the mean value theorem, there exists a point ξ_i such that

$$\sum_j T_{ij} W_j + I_i < \xi_i < \sum_j T_{ij} Y_j + I_i,$$

and

$$\phi_i(W_i) - \phi_i(Y_i) = f_i'(\xi_i)(\sum_{j \neq i} T_{ij}(W_j - Y_j)).$$

Let $\mathbf{T}$ and $\mathbf{M}$ be defined as above. Then, by the triangle inequality

$$\|\phi(W) - \phi(Y)\| \leq \sum_{ij} |f_i'(\xi_i)| \|T_{ij}\| |W_j - Y_j|$$

$$\leq \mathbf{TM} \|W - Y\| = \alpha \|W - Y\|.$$

By hypothesis (ii), $\alpha < 1$, so ϕ is a contraction on S. Then Banach's fixed point theorem implies both the existence of a unique equilibrium W in S and the rate of convergence of the iterates $W(t)$ to W, as in Eq. (4.2). Next, assuming hypothesis (iii) and letting

$$S = \{W : W_i \in \{0, 1\}\},$$

for any fixed $W \in S$, choose

$$I - i = \begin{cases} \theta_1 - \sum_j T_{ij} W_j, & \text{if } W_i - 0, \\ \theta_2 - \sum_j T_{ij} W_j, & \text{if } w_i = 1. \end{cases}$$

Then

$$f_i(\sum_j T_{ij} W_j + I_i) = \begin{cases} f_i(\theta_1) = 0, & \text{if } W_i = 0, \\ f_i(\theta_2) = 1, & \text{if } W_i = 1, \end{cases}$$

or

$$\phi(W) = W.$$

So W is an equilibrium response of the system. Since there are 2^N distinct response vectors in S, this completes the proof of the theorem. $\quad\Box$

Note that the number of steady state responses may be greater than 2^N. Also, Theorem 4.1 does not claim that the state vector $u = (u_i)$ reaches a steady state, but that the respond W does. If the firing rate function is not one-to-one, for instance, if it is flat over some region, then it is possible that the neuron states may not reach equilibrium, although the firing rates do. This is not a major obstacle since we can decide to interpret the recognition of an object as having the response of the system rather than the state itself reach a steady-state.

We would like a condition for stability that is weaker than that given in Theorem 4.1. The principle of linearization tells us that under the conditions we have placed on the nonlinear functions f_i, if the linear part of our dynamical system is stable, then the nonlinear system is also stable. The linear part of the system involves the Jacobian matrix A where each entry (a_{ij}) is given by

$$a_{ij} = T_{ij} f_i'(\sum_{k=1}^{N} T_{ik} W_k^* + I_i).$$

W_k^* is the stable state, $(W_k^*(t + 1) = W_k^*(t))$, where W is as defined above.

The Gershgorin Disc theorem tells us that, with entries $T_{ii} = 0$, all eigenvalues of a matrix lie within a disc D_i where

$$D_i = \{z_i : |z| < \sum_j |a_{ij}|\}.$$

If we place the limit $|a_{ij}| < 1$, so that

$$\forall i \max_{i < i < N} |T_{ik} f_i'(\sum_{k=1}^{N} T_{ik} W_k^* + I_i)| < 1$$

then the eigenvalues of the matrix for the linearized system of (4.1) must lie within the disc with this radius. Since we are restricting it to be less than one, the eigenvalues will be less than one and by the linearity theorem, the matrix is stable.

This is a weaker condition than before since we are taking the maximum over only u_i for which $f_i(u_i)$ is steady, whereas before the maximum M was evaluated over *all* u_i. However, it is also a local result, whereas our previous conditions imply global stability.

An application of Brouwer's Fixed Point Theorem provides another measure of the number of steady states. To see this, first note that $\phi(W)$ is bounded since $W = f(u)$ is restricted to lie between 0 and 1 (or to equal 0 or 1). Here, the unit ball is invariant since $|f(TW + I^i)| \leq 1$. Thus, from Brouwer's theorem, for each of these inputs I^i there is a steady state. Since there are an infinite number of inputs, there would be an infinite number of steady states if each input defines a unique steady state. Thus we seek necessary conditions on the inputs to insure that the steady states are distinct.

Let W^1 be a steady state, and consider two different inputs $I^1 \neq I^2$. If these inputs produce the same steady state response, then we must have

$$f_i(TW^1 + I^1) = f_i(TW^1 + I^2).$$

If $f(u)$ is 1-1, then we have that $I^1 = I^2$. This contradicts our assumption, so that we know that such equality must occur where f is not 1-1, i.e. the places where the derivative of the sigmoid-shaped curve is 0.

Define thresholds θ_1 and θ_2 such that $f_i(u_i) = 0$ for all $u_i < \theta_1$ and $f_i(u_i) = 1$ for all $u_i > \theta_2$. Then, in order for the steady states to be equal, we must have, for at least one i, either both $\sum_{k=1}^{N} T_{ik}W_k^* + I^1 \leq \theta_1$ and $\sum_{k=1}^{N} T_{ik}W_k^* + I^2 \leq \theta_1$ or these sums must be both above θ_2, where W^* is a steady state. Note that it is necessary for this condition to hold for only one i.

Excluding the case where all $u_i < \theta_1$ for one steady state and all $u_i > \theta_2$ for another, if we require the steady states to be different, at least one of the u_i must lie in this interior, 1-1 region of the sigmoid-shaped, activation to firing-rate map. In other words, we must have the condition, for a steady state W^* and for at least one i,

$$\theta_2 > u_i = \sum_k T_{ik}W_k^* + I_i > \theta_1. \tag{4.3}$$

This is not a useful condition since we do not a priori know the steady state W_k^*. Therefore, we expect the worst case and use it to get our condition. Define the norm of a matrix M as $||M|| = \max_i \sum_j |m_{ij}|$. Since $|W|$ is bounded above by 1, $||W|| \leq 1$, where $||W|| = \max_i |W_i|$, and

$$||TW|| \leq ||T|| \cdot ||W|| = ||T||.$$

Thus, two inputs will not result in the same steady state if, for some i,

$$\theta_2 - ||T|| > I_i > \theta_1 + ||T||.$$

This also gives a condition on the thresholds that is necessary for (4.3) to be possible:

$$\theta_2 - \theta_1 > 2||T||.$$

This condition is useful in that it provides some information on the shape of the firing rate curve.

We now turn to the next question that needs to be addressed: how close can two inputs be and still provide distinct responses? We look for a minimum distance between inputs and activation levels which will provide us with distinct steady states. The theorem below gives us bounds on our inputs to insure we have distinct steady states.

We define the distance between two inputs by using the uniform norm:

$$||I^{(1)} - I^{(2)}||_\infty = \max_{1 \le i \le N} |I^{(1)} - I^{(2)}|.$$

We shall drop the subscript "∞" throughout the remainder of this discussion.

Theorem 4.2. *Given a Hopfield-type neural network system, if there exist real numbers $0 < l < L$ such that, for all I^j, I^k,*
  i) $0 < l \le ||I^j - I^k|| \le L$, *and*
  ii) $||T|| \le \alpha l$, $\alpha < 1$;
then

$$L + \alpha l \ge ||U^j - U^k|| \ge l(1 - \alpha),$$

where U^j represents the activation vector resulting from input I^j.

Proof. The distance between two activation levels is

$$||\dot{U}^1 - U^2|| = ||TW^1 + I^1 - TW^2 - I^2|| \tag{4.4}$$
$$= ||I^1 - I^2 + T(W^1 - W^2)||.$$

Because the firing rates W_i are restricted to lie between 0 and 1, $-1 \le W^1 - W^2 \le 1$. Also, by conditions (i) and (ii), the smallest (4.4) can be is $l(1 - \alpha)$. Similarly, the largest (4.4) can be is $L + \alpha l$. This concludes the proof. $\square$

Let us assume that we have a system that meets the above conditions: activation levels are $l(1 - \alpha)$ units apart. We would like to know how many distinguishable outputs there are in such a neural network. We note that activation levels above threshold θ_2 and below θ_1 cannot produce unique steady states since the function is not 1-1 in these regions. The following simple inductive reasoning leads to a lower bound. Consider first a one neuron system. Let us assume that the "interior" region, the 1-1 region between the thresholds, is so small that we can fit in only one activation level. Then we have three distinguishable firing rates: $f^1 = 0$, $f^3 = 1$ and $f^2 = \beta$, where $\theta_1 < \beta < \theta_2$. Now, if we expand this interior region so that it is $l(1 - \alpha)$ wide, we can fit two states in here, and the maximum number of recognizable states increases to 4. Further increasing the width will not allow a greater number of states to be recognized, until the width is $2l(1 - \alpha)$. Now the number is 5. Continuing in this fashion, we see that a lower bound on the maximum number of recognizable states, if all of the above conditions hold, is given by

$$[\frac{\theta_2 - \theta_1}{l(1 - \alpha)}] + 3.$$

The square brackets [ ] indicate that the fraction must be rounded down to the nearest integer.

If we have N neurons, then there are four recognizable states for each neuron. So a lower bound to the total number of recognizable states is

$$\left([\frac{\theta_2 - \theta_1}{l(1 - \alpha)}] + 3\right)^N.$$

4.3 Periodic States

It is easy to construct neural networks that have periodic solutions. We have also already given a condition that prevents periodic behavior. If ϕ in our system is a contraction, that is if $\|\phi(W) - \phi(Y)\| \leq \alpha\|W - Y\|$ where $\alpha < 1$, and W and Y are arbitrary firing rates, then there is a unique steady state and there can be no periodic solutions. In the proof of Theorem 4.1 we showed that a sufficient condition for the system to be a contraction is that $\alpha = \mathbf{MT} < 1$, where

$$\mathbf{M} = \max_{1 \leq i \leq N} |f_i'(u_i)|, \mathbf{T} = \max_{1 \leq j \leq N} \sum_i |T_{ij}|$$

Thus, if the strengths of the synaptic connections are kept small, we will have a contraction.

At this point we note that there is a "folklore" result which claims that if the discrete Hopfield network (4.1) has a symmetric connectivity matrix T, then it has a unique global equilibrium. This result is true for the continuous analogue of system (4.1) as has been proved via the use of a very interesting Liapunov function argument by Grossberg and Cohen [3]. Hopfield [4] has a result for the symmetric discrete system that we are considering here which is often misinterpreted as yielding global stability. His Lyapunov function argument, however, treats the system as if it were a serial, and not a parallel, system. Hopfield considers the system:

$$V_i(t + 1) = 1, \text{ if } \sum_{j \neq i} T_{ij} V_j(t) > U_i$$

$$V_i(t + 1) = 0, \text{ if } \sum_{j \neq i} T_{ij} V_j(t) < U_i.$$

Here, V_i is the firing rate, which we have called f_i, and U_i is the threshold. The thresholds U_i are set equal to 0 as a further simplification. For this system, Hopfield shows that

$$E = -\frac{1}{2} \sum_j \sum_{i \neq j} T_{ij} V_i V_j$$

is a Lyapunov function. This is because the change of energy with respect to time is given by

$$\Delta E = -\Delta V_i \sum_{j \neq i} T_{ij} V_j.$$

ΔE is less than or equal to zero, because if ΔV_i, is negative, the only way this could have come about is if $\sum_{j \neq i} T_{ij} V_j < 0 = U_i$. A similar argument is true for $\Delta V_i > 0$.

The simplification Hopfield has made here is that his system is evolving one neuron at a time. Thus, while the system is spatially parallel, it is temporally serial. This simplification allows Hopfield to keep the V_j term fixed when taking the derivative. As we remarked earlier, there are very simple examples which show that the result does not hold when the network is considered as a parallel system.

We have found conditions on the model (4.1) that assure that it have three of the basic desirable properties for an automatic pattern recognition system. That is, each input leads to a system which has a unique equilibrium, this equilibrium is globally asymptotically stable, and inputs that are sufficiently separated in the uniform norm lead to steady states which are distinguishable. The two remaining questions that need to be treated are the process by which the network would "learn" to recognized objects, and its ability to recognize an object when it is rotated. Clearly, this last concept will lead to an interplay between the uniqueness of equilibria and their distinguishability. We shall next address the question of training the network.

4.4 Unsupervised Learning

Before we start analyzing learning rules we need to define what it is in the model equations that will be affected by such learning. From the broad analogy between biological and model neural networks, it has generally come to be accepted that it is the connectivity matrix T which is to be modified in training the network to do a particular task which, in our case, consists of recognizing objects.

In the theory of Neural Networks there are two general broad classes of methods of "training". The first is called unsupervised learning and consists of a rule by which the connectivity matrix changes under the influence of a sequence of inputs. The second is external or supervised learning which takes several forms, but essentially consists of some type of procedure which adjusts the matrix T in order to optimize a predetermined criterion. It is a form of this second type of learning that was adopted in this problem, but before we describe it, it may be worthwhile to take a fast look at some properties of unsupervised learning rules which essentially make the dynamical system (4.1) non-autonomous. In particular, we can do this by assuming that the system consists of Eq. (4.1) and the learning rule

$$T_{ij}(t) = G(T_{ij}(t-1), \ u(t-1)),$$

even though more general situations where there is dependence on the state of the system at times $t-2, t-3, \cdots, 0$ can also be used. A condition which makes the learning rule unbounded and hence unstable can be easily stated and is an important consideration in developing an adequate unsupervised learning rule.

Theorem 4.3. *Suppose that the connection strength changes only according to its present value and according to the states of the two neurons it connects. Then $T_{ij}(t) = G(T_{ij}(t-1),\ u_i(t),\ u_j(t))$. Suppose there is some synapse such that for some constants $\tilde{u}_i$ and $\tilde{u}_j$ such that $G(w, \tilde{u}_i, \tilde{u}_j) - w$ is a continuous, strictly positive function of w for $w > 0$. Then T_{ij} can become unbounded, and this learning rule is untenable.*

Proof. We can certainly design the input $I(t)$ so that neuron i and neuron j remain at the constant levels $\tilde{u}_i$ and $\tilde{u}_j$ respectively. Then

$$T_{ij}(t) = G(T_{ij}(t-1),\ \tilde{u}_i,\ \tilde{u}_j).$$

Or,

$$T_{ij}(t) = T_{ij}(t-1) + (G(T_{ij}(t-1),\ \tilde{u}_i,\ \tilde{u}_j) - T_{ij}(t-1)).$$

Thus $T_{ij}(t)$ is a strictly increasing function of t. In order to reach a contradiction, assume that $T_{ij}(t)$ is bounded and let W^* be its least upper bound. Then $T_{ij}(t) \to W^*$ as $t \to \infty$. Since $T_{ij}(t)$ is converging, the difference between successive terms is tending to zero. Then $G(w, \tilde{u}_i, \tilde{u}_j) - w \to 0$ as $w \to W^*$. But because G is a continuous function, we must have $G(W^*, \tilde{u}_i, \tilde{u}_j) - W^* = 0$, and this is a contradiction. Therefore $T_{ij}(t)$ becomes unbounded. $\qquad\square$

For example, we consider the learning rule

$$T_{ij}(t) = T_{ij}(t-1) + \eta \frac{(f(u_i(t)) - \frac{1}{2})(f(u_j(t)) - \frac{1}{2})}{1 + T_{ij}^2(t-1)}$$

for some positive constant η. However, for any $\tilde{u}_i$, and $\tilde{u}_j$ both over or both under threshold, this equation becomes

$$T_{ij}(t) = T_{ij}(t-1) + \frac{C}{1 + T_{ij}^2(t-1)},$$

for some positive constant C. Since $\frac{C}{1+T_{ij}^2(t-1)}$ is a strictly positive, continuous function of $T_{ij}(t-1)$, we are able to conclude that the connection strengths can become unbounded, and that this is an undesirable learning rule for our model.

One can also analyze how a neural network would respond when it is not stimulated. This analysis applies to a model which uses our activation equation along with an unspecified learning rule. The following result is quite obvious, given that $f(0) = 0$.

Theorem 4.4. *If $u(t_0) = 0$ and $I(t_0) = 0$ for $t \geq t_0$, then $u(t) = 0$ for $t \geq t_0$.*

One unresolved issue about possible learning rules is whether or not the synapse strength should be changed when neither neuron is firing. We will define connections with *unstimulated learning* to be those for which there does not exist any threshold a such that $T_{ij}(t) = T_{iij}(t-1)$ is true whenever $u_i(t-1) < a$ and $u_j(t-1) < a$. In the absence of this kind of connection, we can make a further claim.

Theorem 4.5. *If the neural network does not have any connections with unstimulated learning, then if $u(t_0) = 0$ and $I(t_0) = 0$ for $t \geq t_0$, then not only do we have $u(t) = 0$ for $t \geq t_0$, but also $T(t) = T(t_0)$ for $t \geq t_0$.*

A modified unsupervised learning rule was implemented in a neural network program called *Athena* that was developed by the team of the first clinic project on this topic. This rule did not have unstimulated learning connections, and it was prevented form generating unbounded connections by a threshold on the absolute value of the synaptic strengths above which the connection remained constant. Several numerical experiments with such rules showed that the network would indeed learn under the stimulus of repeated sequential applications of a collection of different inputs. However, it was found that supervised learning gave a more efficient way of training the network to recognize a given set of objects, and the project concentrated on developing such a learning rule which we will now describe.

4.5 Supervised Learning and Rotational Invariance

In order to simplify the concept of object recognition with rotational invariance we will specialize the neural network in two different ways. The first of these is again motivated by the biological analogy and consists of viewing the neurons as being separated into three groups, commonly called layers, consisting of the input layer, the processing layer and the output layer. The input layer consists of all those neurons (specified by their potentials u_i for $i \in \Omega$) for which the input $I_i \neq 0$, that is, the support of the input I_i. The output layer consists of all those neurons whose output is sampled in determining the state of the system. Thus, instead of sampling the entire state u of the system, or else entire the output $w = f(u)$, one samples a projection of this state

$$O(u) = u\big|_{\Omega_O} \text{ or else } O(f(u)) = f(u)\big|_{\Omega_O},$$

and one defines the output layer to be $\Omega_O \subset \Omega$. The processing layer consists of the neurons in the set obtained by subtracting the supports of $I \in \mathcal{I}$ and Ω_O from Ω. So, the network is layered into three parts, the input layer which can be influenced directly by the external world, the output layer which provides information to the external world, and the hidden or processing layer which connects the first two. This type of structuring of networks into layers was developed by Fukushima [2] who did a large number of computer simulations that yielded promising results for automatic character recognition by neural networks.

The next model that we present is essentially a generalization of the Hopfield network which turned out to be extremely useful in studying rotational invariance and other geometric properties of the input signals. Instead of considering the neurons of the network as being discrete and indexed by the integers $i = 1, 2, \cdots, N$, we assume that they are distributed continuously in a subset Ω of $\mathbf{R}^n$. Thus, instead of neuron potentials $u_i(t)$ at time t, we now have potentials $u(x, t)$ where $x \in \Omega$, and instead of a connectivity matrix T with components

T_{ij} we have a connectivity function T with components $T(x, y)$, $x, y \in \Omega$. The dynamical system (4.1) now generalizes to

$$u(x, t+1) = \int_\Omega T(x, y) f(u(y, t)) dy + I(x), \qquad (4.5)$$

with the steady-states $u(x, t) = \overline{u}(x)$ satisfying

$$\overline{u}(x) = \int_\Omega T(x, y) f(\overline{u}(y, t)) dy + I(x). \qquad (4.6)$$

It is easy to see that the following condition, which is a direct generalization of condition (ii) in Theorem 4.1, is sufficient for the existence and uniqueness of the equilibria given by Eq. (4.6).

$$\max_{x \in [a, b]} \left| \int_\Omega T(x, y) dy \right| \max_{x \in \mathbf{R}} |f'(x)| < 1. \qquad (4.7)$$

Condition (4.7) is also a global stability criterion, hence, when it holds, for any given input I the system will converge to its corresponding unique steady state.

In order to consider rotational invariance, we first define rotations in the standard way as being elements $\mathcal{R}_\alpha$, $\alpha \in [0, 2\pi)$ of a group of invertible linear transformations from $\mathbf{R}^n$ to $\mathbf{R}^n$, with $\det(\mathcal{R}_\alpha) = 1$, $\mathcal{R}_0 =$ identity on $\mathbf{R}^n$, and with

$$\mathcal{R}_\alpha \mathcal{R}_\beta = \mathcal{R}_{\alpha+\beta}.$$

For our particular application, we restrict the rotations $\mathcal{R}_\alpha$ so that $\alpha \in [-\phi, \phi] = [-\pi/12, +\pi/12]$, and define $\| \ \|$ to be an appropriate norm on

$$C = \{f : \overline{\Omega} \to \mathbf{R}, \ f \text{ continuous}\},$$

and $d(A, B)$ to the Hausdorff distance between two sets $A, B \subset C$.

With these definitions we can now give our concept of identification under rotational invariance.

Definition. A neural network of the form (4.5) is δ-*rotational invariant* on $[-\phi, \phi]$ for the collection of inputs $\mathcal{I}$ if it satisfies the following property. Let $\omega(\alpha)$ be the ω limit set of the network with input function $I(\mathcal{R}_\alpha(x))$, then

$$d\big(O(\omega(\alpha)), O(\omega(0))\big) \leq \delta, \quad \forall \ \alpha \in [-\phi, \phi]. \qquad (4.8)$$

When the network has unique steady states, as for example when condition (4.7) holds, then condition (4.8) can be replaced by

$$\|O(\overline{u}(\alpha)) - O(\overline{u}(0))\| \leq \delta, \quad \forall \ \alpha \in [-\phi, \phi], \qquad (4.9)$$

where $\overline{u}(\alpha)$ is the steady-state corresponding to the input $I(\mathcal{R}_\alpha(x))$.

We will say that the neural network has *exact rotational invariance* if $\delta = 0$ in the above definition.

The definition of identification of an image has many possibilities, and after some experimentation the following one was adopted.

Definition. The network (4.5) is said to identify the images in $\mathcal{I}$ if the ω limit set ω_I corresponding to each input $I \in \mathcal{I}$ obeys $O(\omega_I)$ is a unique element of $\mathbf{C}$ with $O(\omega_I) \neq O(\omega_J)$ for any two inputs $I \neq J$ in $\mathcal{I}$, where O is the projection of the states onto the output layer of the network.

This definition does not address the question of separation or resolution of the steady states of distinct images, however, it is clear that one can take care of this by adding a restriction of the form

$$d\big(O(\omega_I), O(\omega_J)\big) > \eta, \tag{4.10}$$

for some appropriate $\eta > 0$, or else the same condition with some functional replacing the norm in Eq. (4.10).

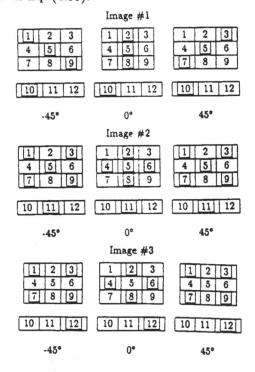

Fig. 4.1. Three discretized images in three rotations

These definitions allow us to approach the problem of rotational invariance as an optimization problem in which the connectivity function T is chosen so as to minimize δ in relation (4.9) while maximizing

$$\sum_{(I,J),\ I \neq J} d\big(O(\omega_I), O(\omega_J)\big), \quad I, J \in \mathcal{I}.$$

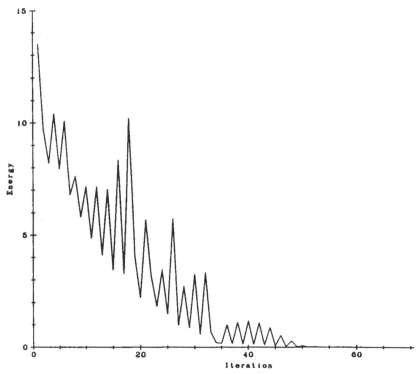

Fig. 4.2. Variation of the energy ψ_{total} with iteration

The difficulty with this approach is that the omega limit sets involved in the above optimization criterion are difficult to compute. An alternate method that is computationally more feasible is suggested by the following result.

Theorem 4.6. *Let $\Omega_O \subset \Omega$ be the observable layer of the neural network (4.5), and suppose that the network has a unique globally attractive steady-state $\overline{u}_\alpha$ corresponding to the input I_α, where $I_\alpha(x) = I(\mathcal{R}_\alpha(x))$. Then the network has exact rotational invariance on the input set $\mathcal{I}$ if for each $I \in \mathcal{I}$ and each $\alpha,\ \beta \in [-\phi, \phi]$ we have*

$$\int_{\Omega - \Omega_O} T(x,y)[f(\overline{u}_\alpha(y)) - f(\overline{u}_\beta(y))]dy = 0, \quad \forall x \in \Omega_O.$$

The proof of this result follows directly from Eq. (4.6) for the steady-states and the definition of exact rotational invariance. It suggests that in the optimization procedure for rotational invariance one needs to minimize the *total rotational error*

$$\psi_{\text{rotate}} = \int_{-\phi}^{\phi} \int_{-\phi}^{\phi} \Big[\int_{\Omega - \Omega_O} T(x,y)[f(\overline{u}_\alpha(y)) - f(\overline{u}_\beta(y))]dy \Big]^2 d\alpha d\beta. \qquad (4.11)$$

The unique identification optimization problem is handled as follows. Assumed that the set of images $\mathcal{I}$ is finite, and denote the unique steady-state of the ith image in $\mathcal{I}$ by $\overline{u}^i$, $i = 1, 2, \cdots, n$. Associate with the ith input a subset $\Omega_O^i \subset \Omega_O$ of measure $|\Omega_O|/n$, and such that $|\Omega_O^i \cap \Omega_O^j| = 0$ for $i \neq j$. Here $|\Omega|$ denotes the n-dimensional Lebesgue measure of Ω. The identification of an image is then associated with maximizing the output in its associated region while minimizing it in its complement in Ω_O. The maximum can be arbitrarily chosen, and here we set it equal to one and define

$$
\psi_{\text{identify}} = \sum_{i=1}^{i=n} \Big\{ \int_{-\phi}^{\phi} \int_{\Omega_O^i} \big[1 - u_\alpha^i(y) \big]^2 dy d\alpha
$$
$$
+ \sum_{j \neq i} \int_{-\phi}^{\phi} \int_{\Omega_O^j} u_\alpha^j(y)^2 dy d\alpha \Big\}.
\tag{4.12}
$$

The optimization criterion is to minimize

$$
\psi_{\text{total}} = \psi_{\text{rotate}} + c \psi_{\text{identify}}
$$

where c is a weight which allows variation in the relative emphasis that is placed on identification versus rotational invariance.

In implementing the above optimization criterion the connectivity function T was represented by a finite sum of basis functions $t_{i,j}$ and a variety of different standard finite dimensional optimization methods were used. A numerical example of this method that was computed consisted of a neural network with nine input neurons and three output neurons which was trained to recognize the three images each in three rotations that are depicted in Fig. 4.1 together with the corresponding identification subregion in the output layer for each of the images. For each of these images the inputs to the first nine neurons corresponding to the numbers that are framed in Fig. 4.1 were set equal to one and the others were set equal to zero. Similarly, the target response of the three output layer neuron for each image equals to one for the neuron that is framed and zero for the two others.

A steepest descent algorithm was used to minimize ψ_{total} and the graph of the change of this energy with the iteration is shown in Fig. 4.2. It is clear from this that by the fiftieth iteration the procedure has converged.

Once the connectivity function T of this network is optimized it recognizes these three images in the sense that the output layer neurons have responses that correspond to the target response when any one of the rotated versions of the image is presented to the network. This way of selecting the connectivity strengths T falls within the scope of supervised learning, since the target outputs are preselected and an external procedure is used to optimize the choice of T.

4.6 Ending Note

This particular project was the basis of substantial further work by the sponsor. In fact, one of the students, Forrest Armstrong, who was on the first and second clinic team that worked on the project was hired by General Dynamics to continue developing this area. This work, which included a third Mathematics Clinic project with Mr. Armstrong in the industrial liaison role this time, led to semiconductor implementations of image classification neural networks and to the recent establishment of a small independent branch of the company that is producing these neural networks designs. There has been a transfer of technology because of this project in the other direction as well, from industry to the university. In fact, the interest of the students and of the faculty that was stimulated by this work led to the establishment of a course in Neural Networks which is now part of the computer science curriculum at Harvey Mudd College.

References

1. Armstrong, F., J. Brennock, K. Ring, and L. Rossi (student team), S. Busenberg (faculty), R. Schlunt (liaison), H. Thieme (consultant), (1988): Mathematical Modeling and Simulation of a Neural Image Classifier, Harvey Mudd College Mathematics Clinic Report.
2. Fukushima, K., (1980): Neocognitron: a self-organizing network model for a mechanism of pattern recognition unaffected by shift in position, *Biological Cybernetics*, **36**, pp. 193-202.
3. Grossberg, S. and M. Cohen (1983): Stability of global pattern formation and parallel memory storage by competitive neural networks, *IEEE Trans. on Systems, Man, and Cybernetics*, **13**, pp. 815-826.
4. Hopfield, J. (1982): Neural networks and Physical systems with emergent collective computational abilities, *Proc. Natl. Acad. Sci. USA*, **79**, pp. 2554-2558.

List of Participants

C. ALDRIDGE, Math. Inst., 24-29 St. Giles, Oxford, OX11 3LB

D. ANDREUCCI, Dip. di Mat., Viale Morgagni 67/A, 50134 Firenze

L. ARLOTTI, Dip. di Mat., Fac. di Ing., Via Brecce Bianche, 60100 Ancona

L. BORRELLI, Via Luigi Milella 4, 70124 Bari

S. BUSENBERG, Dept. of Math., Harvey Mudd College, Claremont, CA 91711

G. BUSONI, Dip. di Mat., Viale Morgagni 67/A, 50134 Firenze

S. CANDIA, Via Fiume 10, Monopoli, Bari

V. CAPASSO, Director SASIAM, Tecnopolis, 70010 Valenzano, Bari

R. CASELLI, SASIAM, Tecnopolis, 70010 Valenzano, Bari

R. CONTI, Dip. di Mat. Appl., Via S. Marta 3, 50139 Firenze

A. DI LIDDO, IRMA/CNR, Dip. di Mat., Campus Universitario, Via Fortunato, 70125 Bari

J. DYLANDER, Mat. Inst., Universitetsparken 5, 2100 Copenhagen

A. FASANO, Dip. di Mat., Viale Morgagni 67/A, 50134 Firenze

B. FORTE, Dept. of Appl. Math., Univ. of Waterloo, Waterloo, Ontario, Canada N2L 3G1

C.-I. GHEORGHIU, Inst. de Mat., Oficiul Postal nr.1, C.P. 68, 3400 Cluj-Napoca

R. GIANNI, Dip. di Mat., Viale Morgagni 67/A, 50134 Firenze

M. HEILIO, Dept. of Inf. Techn., Lappeenranta Univ. of Techn., 53851 Lappeenranta

C. HUNTINGFORD, Hertford College M.C.R., Oxford, OX1 3BW

H. K. KUIKEN, Philips Res. Lab., P.O. Box 80000, 5600 JA Eindhoven

P. LI, c/o T. Gianfrancesco, Via Papa Innocenzo XII n. 8, 70124 Bari

E. LOPANE, Via Corfù 10, 70100 Bari

P. MANNUCCI, Dip. di Mat., Viale Morgagni 67/A, 50134 Firenze

S. MARMI, Dip. di Mat., Viale Morgagni 67/A, 50134 Firenze

R. MININNI, Via Principe Amedeo 26, 70100 Bari

M. OLESEN, Mat. Inst., Universitetsparken 5, 2100 Copenhagen

J. OTTESEN, Inst. of Math. and Phys., Roskilde Universitycentre, Postbox 260,
   DK-4000 Roskilde

M. PRIMICERIO, Dip. di Mat., Viale Morgagni 67/A, 50134 Firenze

A. REPACI, Dip. di Mat., Politecnico, Corso Duca degli Abruzzi 24; 10129 Torino

R. RICCI, Dip. di Mat., Viale Morgagni 67/A, 50134 Firenze

H. THYGESEN, Mat. Inst., Universitetsparken 5, 2100 Copenhagen

U. TOLVE, SASIAM, Tecnopolis, 70010 Valenzano, Bari

A. TORELLI, Dip. di Mat., Strada Nuova 65, 27100 Pavia

FONDAZIONE C.I.M.E.
CENTRO INTERNAZIONALE MATEMATICO ESTIVO
INTERNATIONAL MATHEMATICAL SUMMER CENTER

"Topological Methods in the Theory of Ordinary Differential Equations in Finite and Infinite Dimensions"

is the subject of the First 1991 C.I.M.E. Session.

The Session, sponsored by the Consiglio Nazionale delle Ricerche and the Ministero dell'Università e della Ricerca Scientifica e Tecnologica, will take place under the scientific direction of Prof. MASSIMO FURI (Università di Firenze), and Prof. PIETRO ZECCA (Università di Firenze) at Villa "La Querceta", Montecatini (Pistoia), **from June 24 to July 2, 1991.**

Courses

a) **Quasilinear Evolution Equations of Parabolic Type.** (5 lectures in English).
   Prof. Herbert AMANN (Universität Zürich).

Outline

Abstract quasilinear parabolic evolution equations in Banach spaces form a convenient framework for the study of a variety of concrete quasilinear parabolic systems of reaction-diffusion type occuring in applications. By means of interpolation techniques it can be shown that they generate local semiflows on suitable Banach spaces. Given appropriate assumptions, these semiflows can be analyzed further to deduce information about global existence, blow up, stability of critical points, bifurcations of periodic orbits, etc. These questions will be discussed and applications to concrete systems will be given.

References

H. Amann: Dynamic theory of quasilinear parabolic systems.
 I. Abstract evolution equations, Nonl. Anal. T.M. & A. 12 (1988), 895-919
 II. Reaction-diffusion systems, Diff. Int. Eqns. 3 (1990), 13-75.
 III. Global existence, Math. Z. 202 (1989), 219-250. Erratum, Math. Z. 205 (1990), 331.
H. Amann: Hopf bifurcation in quasilinear reaction-diffusion systems. Proc. Claremont Conf. Diff. Eqns. Appl. to Biology and Pop. Dynamics 1990.
G. Da Prato & A. Lunardi: Hopf bifurcation for fully nonlinear equations in Banach space, Ann. Inst. Henri Poincaré - Analyse non linéaire 3 (1986), 315-329.
J. Escher: Global existence and nonexistence for semilinear parabolic systems with nonlinear boundary conditions, Math. Annalen, 284 (1989), 285-306.

b) **Families of Nonlinear Fredholm Operators.** (5 lectures in English).
   Prof. Patrick FITZPATRICK (University of Maryland).

Lecture 1 - Topology of Spaces of Linear Operators and the Implications for Degree Theory.
Lecture 2 - The Parity as a Bifurcation Invariant.
Lecture 3 - Degree Theory and Nonlinear Elliptic Boundary Value Problems.
Lecture 4 - Degree Theory for C² Fredholm Mappings
Lecture 5 - The Analytical Index and Two-Point Boundary Value Problems.

References

1. M. Atiyah, K-Theory, W.A. Benjamin, New York, 1967.
2. F.E. Browder, Nonlinear Operators and Nonlinear Equations of Evolution in Banach Spaces, Proc. Symp. Pure Math., Vol. 18, Part 2 (1976).
3. R. Caccioppoli, Sulle corrispondenze funzionali inverse diramate: teoria generale e applicazioni ad alcune equazioni funzionali non lineari, e al problema di Plateau, Opere Scelte, Vol. II, Edizioni Cremonese, Roma, 1963, pp. 157-177.
4. Shiu-Nee Chow and Jack K. Hale, Methods of Bifurcation Theory, Grund. der mat. Wissen., Vol. 251 (1982), Springer Verlag.
5. K.D. Elworthy and A.J. Tromba, Degree Theory on Banach Manifolds, in Global Analysis, Proc. Symp. Pure Math., Vol. 15 (1970), pp. 45-94.
6. C.C. Fenske, Extensio gradus ad quasdam applicationes Fredholmii, Mitt. Mat. Sem. Giessen, Heft 121 (1976), pp. 65-70.
7. P.M. Fitzpatrick, The stability of parity and global bifurcation via Galerkin approximation, J. London Math. Soc., 2(38), (1988), pp. 153-165
8. P.M. Fitzpatrick, and J. Pejsachowicz, An extension of the Leray-Schauder degree for fully nonlinear elliptic problems, Proc. Symp. Pure Math. 45, Part 1 (1986), pp. 425-439.
9. P.M. Fitzpatrick and J. Pejsachowicz, The fundamental group of the space of linear Fredholm operators and the global analysis of semilinear equations, Contemporary Math., Vol. 12 (1988), pp. 47-87.
10. P.M. Fitzpatrick and J. Pejsachowicz, A local bifurcation theorem for C^1- Fredholm maps, Proc. of the Amer. Math. Soc., Vol. 109, 4 (1990), pp. 995-1002.
11. P.M. Fitzpatrick and J. Pejsachowicz, Parity and generalized multiplicity, Trans. of the Amer. Math. Soc. (in press).
12. P.M. Fitzpatrick and J. Pejsachowicz, Nonorientability of the index bundle and several-parameter bifurcation, J. of Functional Anal. (in press).
13. P.M. Fitzpatrick and J. Pejsachowicz, Orientation and the Leray-Schauder theory for fully non linear boundary value problems, Memoirs of the American Math. Soc. (in press).
14. P.M. Fitzpatrick, J. Pejsachowicz and P. Rabier, Topological degree for nonlinear Fredholm operators, C.R. Acad. Sci. Paris (1990), t. 311, Serie I, pp. 711-716.
15. J. Ize, Bifurcation theory for Fredholm operators, Memoirs of the Amer. Math. Soc. No. 175 (1975).
16. J. Ize, Necessary and sufficient conditions for multiparameter bifurcation, Rocky Mountain J. of Math., vol. 18 (1988), pp. 305-337.
17. T. Kato, Perturbation theory for linear operators, 2nd ed., Grund. der mat. Wissen., vol. 132, Springer Verlag, 1980.
18. H. Kielhofer, Multiple eigenvalue bifurcation for Fredholm operators, J. fur die reine und angewandte Mat., Band 358 (1985), pp. 104-124.
19. M.A. Krasnosel'skii and P.P. Zabreiko, Geometrical methods in nonlinear analysis, Grund. der mat. Wissen., Vol. 263 (1984), Springer Verlag.
20. J. Leray and J. Schauder, Topologie et équations fonctionnelles, Ann. Sci. Ecole Norm. Sup. 51 (1934), pp. 45-78.
21. Yan Yan Li, Degree of second order nonlinear elliptic operators and its applications, Comm. in Part. Diff. Eq. 14 (11), (1989), pp. 1541-1571.
22. J. Mawhin, Topological degree methods in nonlinear value problems, Conference Board of the Mathematical Sciences, No. 40 (1977), Amer. Math. Soc.
23. B.S. Mitijagin, The homotopy structure of the linear group of a Banach space, Uspeki Mat. Nauk, Vol. 72 (1970), pp. 63-106.
24. L. Nirenberg, Variational and topological methods in nonlinear problems, Bull. Amer. Math. Soc. (N.S.) 4 (1981), pp. 267-302.
25. J. Pejsachowicz and A. Vignoli, On the topological concidence degree of perturbation of Fredholm operator operator, Boll. U.M.I., (5), (1980), pp. 1457-1466.
26. F. Quinn and A. Sard, Hausdorff conullity of critical images of Fredholm maps, Amer. J. of Math., 94 (1972), pp. 1101-1110.
27. A.I. Snirel'mann, The degree of a quasi-ruled mapping and a nonlinear Hilbert problem, Math. USSR-Sb. 18 (1972), pp. 373-396.
28. M.G. Zaidenberg, S.G. Krein, P.A. Kuchment and A.A. Pankov, Banach bundles and linear operators, Russian Math. Surveys, 30 5 (1975), pp. 115-175.

c) **Continuation Principles and Boundary Value Problems.** (5 lectures in English).

Prof. Mario MARTELLI (California State University at Fullerton).

Outline

The course will illustrate how the basic idea of continuation has been used to obtain existence results and, sometimes, to constructively produce solutions of non linear operators equations derived particularly, but not exclusively, from boundary value problems. Bifurcation and cobifurcation settings will be analyzed. Attempts at extending continuation methods to critical point theory will also be discussed.

References

A. Granas, The theory of compact vector fields and some of its applications to the topology of functional spaces, Rozprawy Mathematiczne, Warsaw, 30, (1962).
J. Mawhin, Functional analysis and boundary value problems, Studies in ordinary differential equations, J. Hale ed., MAA Studies in mathematics, 14 (1977), 128-168.
G. Prodi, A. Ambrosetti, Analisi non lineare, Quaderni della Scuola Normale Superiore di Pisa, I Quaderno, 1973.
P. Rabinowitz, Théorie du degré topologique et applications à des problèmes aux limites non linéaires, Université Paris VI, (1975).
M. Furi, M. Martelli, A. Vignoli, Contribution to the spectral theory of nonlinear operators in Banach Spaces, Annali di Matematica Pura e Applicata, 118 (1978), 229-294.

d) **Topological Degree and Boundary Value Problems for Nonlinear Differential Equations.** (5 lectures in English).
   Prof. Jean MAWHIN (Université Catholique de Louvain).

Outline

Significant contributions in the use of topological degree methods in the study of nonlinear boundary value problems for ordinary differential equations have been made in the recent years. They rely upon new theoretical results in basic degree theory and on new techniques for obtaining a priori estimates for the solutions of differential equations. Special emphasis will be put on the following aspects.

1. The use of equivariance properties in the computation of the degree of fixed points operators associated to periodic problems for ordinary differential equations.

2. Superlinear differential equations and systems.

3. Asymptotically positive homogeneous differential systems.

4. Recent results in nonresonant and resonant nonlinear boundary value problems.

5. Some new methods for the a priori estimates of solutions of nonlinear boundary value problems.

6. The use of topological degree in the study of the stability of periodic solutions.

References

J. Mawhin, Topological degree methods in nonlinear boundary value problems. CBMS Regional Conference Series in Mathematics, Nr 40. American Math. Soc., Providence, 1977.

M.A. Krasnosel'skii and P.P. Zabreiko. Geometrical methods of nonlinear analysis. Grundlehren der mathematische Wissenschaften, Nr 263. Springer, Berlin, 1984.

A. Capietto, J. Mawhin and F. Zanolin, A continuation approach to superlinear boundary value problems, J. Differential Equations, 88 (1990), 347-395.

A. Capietto, J. Mawhin and F. Zanolin. Continuation theorems for periodic perturbations of autonomous systems, Institut de math. pure et appl., Univ. Cath. Louvain, Rapport n. 150, 1989. To appear in Trans. Amer. Math. Soc.

Th. Bartsch and J. Mawhin, The Leray-Schauder degree of S^1-equivariant operators associated to autonomous neutral equations in spaces of periodic functions. Institut de math. pure et appl., Univ. Cath. Louvain, Rapport n. 184, 1990. To appear in J. Differential Equations.

R. Ortega, Stability and index of periodic solutions of an equation of Duffing type, Bollettino Un. Mat. Ital. (7) 3-B (1089), 533-546.

R. Ortega, Stability of a periodic problem of Ambrosetti-Prodi type, Differential and Integral Equations, 3 (1990), 275-284

e) **Some Applications of the Fixed Point Index in Analysis.** (5 lectures).
   Prof. Roger NUSSBAUM (Rutgers University).

Outline

The fixed point index is a generalization of both the Leray-Schauder degree and the Lefschetz fixed point theorem. This course will describe the basic properties of the fixed point index and sketch proofs of some important theorems, notably the so-called map theorem of Krasnosel'skii, Zabreiko and Steinlein. Applications will be given to nonlinear differential-delay equations, in particular, to some recent unpublished work of Mallet-Paret and Nussbaum on differential-delay equations with state dependent time lags.

References

1. Robert F. Brown, The Lefschetz Fixed Point Theorem, Scott, Foresman and Co., Glenview, Illinois, 1971.

2. Roger D. Nussbaum. The fixed point index and some applications, Lecture notes, Seminaire de Mathématiques supérieures, Les Presses de l'Université de Montréal, 94 (1985), 1-145.

3. R.D. Nussbaum and J. Mallet-Paret, "Global continuation and asymptotic behaviour for periodic solutions of a differential-delay equations", Annali di Matematica pura ed appl., 145 (1986), 33- 128.

4. J. Mallet-Paret and R.D. Nussbaum, "A differential-delay equation arising in optics and physiology", SIAM J. Math. Analysis, 20 (1989), 249-292.

LIST OF C.I.M.E. SEMINARS   Publisher

1963 - 29. Equazioni differenziali astratte   "

       30. Funzioni e varietà complesse   "

       31. Proprietà di media e teoremi di confronto in Fisica Matematica   "

1964 - 32. Relatività generale   "

       33. Dinamica dei gas rarefatti   "

       34. Alcune questioni di analisi numerica   "

       35. Equazioni differenziali non lineari   "

1965 - 36. Non-linear continuum theories   "

       37. Some aspects of ring theory   "

       38. Mathematical optimization in economics   "

1966 - 39. Calculus of variations   Ed. Cremonese, Firenze

       40. Economia matematica   "

       41. Classi caratteristiche e questioni connesse   "

       42. Some aspects of diffusion theory   "

1967 - 43. Modern questions of celestial mechanics   "

       44. Numerical analysis of partial differential equations   "

       45. Geometry of homogeneous bounded domains   "

1968 - 46. Controllability and observability   "

       47. Pseudo-differential operators   "

       48. Aspects of mathematical logic   "

1969 - 49. Potential theory   "

       50. Non-linear continuum theories in mechanics and physics

          and their applications   "

       51. Questions of algebraic varieties   "

1970 - 52. Relativistic fluid dynamics   "

       53. Theory of group representations and Fourier analysis   "

       54. Functional equations and inequalities   "

       55. Problems in non-linear analysis   "

1971 - 56. Stereodynamics   "

       57. Constructive aspects of functional analysis (2 vol.)   "

       58. Categories and commutative algebra   "

1972 - 59. Non-linear mechanics   "
        60. Finite geometric structures and their applications   "
        61. Geometric measure theory and minimal surfaces   "

1973 - 62. Complex analysis   "
        63. New variational techniques in mathematical physics   "
        64. Spectral analysis   "

1974 - 65. Stability problems   "
        66. Singularities of analytic spaces   "
        67. Eigenvalues of non linear problems   "

1975 - 68. Theoretical computer sciences   "
        69. Model theory and applications   "
        70. Differential operators and manifolds   "

1976 - 71. Statistical Mechanics   Ed Liguori, Napoli
        72. Hyperbolicity   "
        73. Differential topology   "

1977 - 74. Materials with memory   "
        75. Pseudodifferential operators with applications   "
        76. Algebraic surfaces   "

1978 - 77. Stochastic differential equations   "
        78. Dynamical systems   Ed Liguori, Napoli and Birhäuser Verlag

1979 - 79. Recursion theory and computational complexity   "
        80. Mathematics of biology   "

1980 - 81. Wave propagation   "
        82. Harmonic analysis and group representations   "
        83. Matroid theory and its applications   "

1981 - 84. Kinetic Theories and the Boltzmann Equation (LNM 1048) Springer-Verlag
        85. Algebraic Threefolds   (LNM 947)   "
        86. Nonlinear Filtering and Stochastic Control (LNM 972)   "

1982 - 87. Invariant Theory   (LNM 996)   "
        88. Thermodynamics and Constitutive Equations (LN Physics 228)   "
        89. Fluid Dynamics   (LNM 1047)   "

1983 - 90. Complete Intersections   (LNM 1092)   "
      91. Bifurcation Theory and Applications   (LNM 1057)   "
      92. Numerical Methods in Fluid Dynamics   (LNM 1127)   "

1984 - 93. Harmonic Mappings and Minimal Immersions   (LNM 1161)   "
      94. Schrödinger Operators   (LNM 1159)   "
      95. Buildings and the Geometry of Diagrams   (LNM 1181)   "

1985 - 96. Probability and Analysis   (LNM 1206)   "
      97. Some Problems in Nonlinear Diffusion   (LNM 1224)   "
      98. Theory of Moduli   (LNM 1337)   "

1986 - 99. Inverse Problems   (LNM 1225)   "
     100. Mathematical Economics   (LNM 1330)   "
     101. Combinatorial Optimization   (LNM 1403)   "

1987 - 102. Relativistic Fluid Dynamics   (LNM 1385)   "
     103. Topics in Calculus of Variations   (LNM 1365)   "

1988 - 104. Logic and Computer Science   (LNM 1429)   "
     105. Global Geometry and Mathematical Physics (LNM 1451)   "

1989 - 106. Methods of nonconvex analysis   (LNM 1446)   "
     107. Microlocal Analysis and Applications   (LNM 1495)   "

1990 - 108. Geoemtric Topology: Recent Developments   (LNM 1504)   "
     109. H$_\infty$ Control Theory   (LNM 1496)   "
     110. Mathematical Modelling of Industrical   (LNM 1521)   "
         Processes

1991 - 111. Topological Methods in the Theory of   to appear   "
         Ordinary Differential Equations in Finite
         and Infinite Dimensions
     112. Arithmetic Algebraic Geometry   to appear   "
     113. Transition to Chaos in Classical and   to appear   "
         Quantum Mechanics

Lecture Notes in Mathematics

For information about Vols. 1–1323
please contact your bookseller or Springer-Verlag